GOD, CHANCE & NECESSITY

'At last, God is beginning to argue back.'

Clive Cookson in the *Financial Times*

'A witty, clear and probing critique of the atheist reductionists . . . This is a lively and important book . . . The debate between theism and materialism is of the highest significance and Keith Ward has made a worthy contribution to it.'

John Polkinghorne in the
Times Higher Educational Supplement

'Beautifully and clearly written, accessible.'

Peter Atkins in *The Observer*

'What is fascinating and new about the book is Ward's unabashed animus and determination to take the fight to the scientists.'

Henry Porter in *The Guardian*

'A profound and richly satisfying book that illuminates better than anything I have come across the most important issue of our time and which sooner or later will consign atheism to the dustbin of philosophical ideas.'

James le Fanu in the *Catholic Herald*

'This

Other books by Keith Ward

The Concept of God
Defending the Soul
Holding Fast to God
Images of Eternity
Religion and Creation
Religion and Revelation
The Turn of the Tide
A Vision to Pursue

GOD, CHANCE & NECESSITY

....

KEITH WARD

ONEWORLD
OXFORD

GOD, CHANCE AND NECESSITY

Oneworld Publications
(Sales and Editorial)
185 Banbury Road
Oxford OX2 7AR
England

Oneworld Publications
(U.S. Marketing Office)
PO Box 830, 21 Broadway
Rockport, MA 01966
U.S.A.

ISBN 1–85168–116–7

Cover design and illustration by Peter Maguire
Printed and bound in Finland by WSOY

Contents

Acknowledgements

I am extremely grateful to Dr Arthur Peacocke and Dr Peter Hodgson of the University of Oxford, who kindly read parts of the manuscript and made many very valuable suggestions. I am also grateful to John Polkinghorne, Chris Isham and the members of the 1991 Vatican Observatory Research Conference, who have helped to shape my understanding of contemporary science.

Introduction

Bereshit barah Elohim: 'In the beginning, God created the heavens and the earth.' (Gen. 1.1) To the majority of those who have reflected deeply and written about the origin and nature of the universe, it has seemed that it points beyond itself to a source which is non-physical and of great intelligence and power. Almost all the great classical philosophers – certainly Plato, Aristotle, Descartes, Leibniz, Spinoza, Kant, Hegel, Locke, Berkeley – saw the origin of the universe as lying in a transcendent reality. They had different specific ideas of this reality, and different ways of approaching it, but that the universe is not self-explanatory and that it requires some explanation beyond itself was something they accepted as fairly obvious.

Religious thinkers, too, have usually seen the object of their worship as a creator God who brings the universe into existence, so that the universe is not an independent reality, but comes from a spiritual source beyond it. This is obviously true of Semitic religions like Judaism, Christianity and Islam, but it is true also of most Indian traditions, which, at least at a reflective level, think of one supreme spiritual source under many different names.

Of course there is no universal agreement on these issues. Philosophers like David Hume and religions like Buddhism have denied the usefulness and even the coherence of the idea of a God. As one surveys the history of human thought, it seems that there is a major division between those who see the fundamental reality as spiritual, purposive and of supreme value, and those who see it as material and without any purpose or value, except what individual humans may choose to give it. The former group, the theists, are certainly in a majority, historically speaking, but counting heads is no criterion of truth. Moreover, some would claim that the majority belongs to the past, before the rise of modern science, and that the situation has now changed. In the twentieth century especially, there has arisen a new and self-confident movement which asserts that theism is outdated, that any theory of creation is unnecessary, and that scientific truth is incompatible with religious faith. This movement may in general and not unfairly be called 'evolutionary naturalism'.[1]

After Nietzsche preached, in the name of human freedom, that God was dead, and Auguste Comte declared that positivistic (empirically based) science had supplanted religious mythology, the way was open for atheists who happened to be scientists to say that it was science as such that conflicted with religion. It is not very easy for any scientist to do this with complete honesty, since all scientists have well-respected colleagues, with Nobel prizes and scientific honours of all sorts, who have deep religious convictions. It is one thing to say that one disagrees with colleagues. It is quite another thing to say that they are blind, ignorant and deluded, since they are unable to see that the science they practise excludes the faith they also superstitiously practise. I suppose that in this the atheists are just taking revenge for the assertions believers sometimes make that religious unbelief is the result of ignorance and delusion. Nevertheless, the spectacle of two groups of people, representing the greatest intellectual resources of the planet, calling each other deluded and ignorant, is hardly an edifying one. I have no intention of contributing to it. I think atheists are mistaken, but I do not think they are either deluded (i.e. duped into believing something) or ignorant.

THE SCIENTIFIC WORLDVIEW

It is certainly true that the intellectual scene has changed because of the rise of the natural sciences. The main way in which it has changed is that we now have a view of the universe as a vast expanse, containing millions of galaxies and governed by universal laws which can be mathematically described and ordered in one coherent system. In the twentieth century, the structure of the atom has been discovered, allowing the fundamental elements of matter to be categorised. The structure of the nucleic acids that form the basis of life has been decoded, so that we can see the mechanisms of heredity in some detail. The brain is being mapped, so that we can see how perception and thought work, and can identify the states of the brain associated with perception and thought. Each year, we achieve a better understanding of the physical basis of our existence and of the world around us. We can see how these things work, and we can begin to manipulate them according to our own designs.

The achievements of the sciences are truly amazing, so amazing that they sometimes lead to a belief that humans are now the sole masters of the universe, able to manipulate it as they will. For some ancient Greek philosophers, humanity was the measure of all things, in theory. Now, however, humanity can be the measure of all things in fact as well.

This thought is not an entirely happy one if the applications of scientific discovery are not carefully guided by some system of values. Humans have used the discoveries of nuclear physicists to devise bombs that can eliminate all life on earth, to generate viruses that can destroy living organisms, and to pollute the environment with chemicals so that it may not support life for very much longer anyway. The only trouble with gaining power over our lives is that it is people like us who have the power. And if our personal lives are anything to go by, that power will be used for terribly destructive ends. To that extent, the growth of science is an ambiguous good.

That is not an argument for stopping scientific investigation. But it does raise a warning against any easy assumption that all science is good, that there are no limits on the drive to experimentation, and that humans are free to manipulate the universe at will. Far from being masters of the universe, humans are not even masters of themselves. It may seem liberating to proclaim freedom from tradition and constraint. Nietzsche certainly thought it was. Unfortunately, anything which encourages the will to power in the modern world is actually encouraging the destruction of the earth.

Of course, atheists can see that as well as anyone else. But it is clear that the fashionable call to freedom from any idea of an objective purpose and value in the universe can have undesirable consequences. One needs to be very careful about asking if modern science really does commit one to rejecting objective purposes and values. As a matter of fact, the scientific attitude is deeply imbued with one major value at least, the value of seeking truth for its own sake. Regrettably, some atheistic scientists have a very restricted view of what truth is, and of what may lead to its discovery. They think that truth lies only in what can be measured and experimentally tested, and that the only way to discover it is by cool analysis and dispassionate observation. This is the programme of naturalism. Thus arises a kind of scientistic barbarism, which sees the study of the humanities, of literature, philosophy, history and art, as useless time-wasting. Technology and manipulation thrive, whereas reflection and contemplation wither. The completely technocratic and amoral society emerges, able to manipulate anything but appreciate nothing; a desert of the mind.

It does not have to be like that. Science has its own proper form of contemplation, of reverence and awe before the beauty and wisdom of the natural world, of trust in a community of scholarship and understanding, of openness to wider visions of truth. In a way, scientists have a great advantage over philosophers and theologians. In addition to being numerate, and being able to understand the intricacies of nature, they can also appreciate art, morality and reflection. The truly scientific mind can also be open to forms of truth which lie in the area of personal life and relationships. Some truths can only be approached by personal commitment and involvement, and knowledge of them transforms the knower, by an essentially individual (but not just subjective) interaction.[2] Religious truth is of that kind, and to ignore it altogether is to ignore a vast dimension of human experience. One would expect good scientists to be more open to these wider dimensions of truth than those who are not so aware of the awe-inspiring structure of the world. I think that, in the main, they are. Great scientists, like Newton, Faraday, Maxwell and Einstein, may not always belong to any completely orthodox religious tradition, but they very often show an awareness of a vast intelligence underlying the universe, and a reverence before the mystery of existence.

THE NEW MATERIALISM

Regrettably, a form of materialism which is entirely hostile to religion, and which mocks any idea of objective purpose and value in the universe, has become fashionable in recent years. Good scientists like Francis Crick, Carl Sagan, Stephen Hawking, Richard Dawkins, Jacques Monod and Peter Atkins have published books that openly deride religious beliefs, and claim the authority of their own scientific work for their attacks. Their claims are seriously misplaced. Their properly scientific work has no particular relevance to the truth or falsity of most religious claims. When they do stray into the fields of philosophy, they ignore both the history and the diversity of philosophical viewpoints, pretending that materialist views are almost universally held, when, in fact, they are held by only a fairly small minority among philosophers ('theologian', of course, is for them only a term of abuse). The form of materialism they espouse is open to very strong, and standard, criticisms, particularly in respect to its virtual total inability to account for the facts of consciousness and for the importance of ideas of truth and virtue. These scientists rarely disclose how very contentious their views are among their own colleagues, or how little we actually know about the subjects upon which they write so confidently. Ironically, their attitudes are often anti-scientific in temper as well as anti-religious, since they do not consider carefully and rigorously the claims of major theologians, but are content to lampoon the crudest versions of the most naive religious doctrines they can find. Their treatment of religion shows no dispassionate analysis, but a virulent contempt which can only be termed prejudice.

It may seem pointless to attempt a reasoned response to collections of misplaced prejudices. The fact is, however, that all these writers are good scientists and brilliant writers. Their books are certainly worth reading, and can be commended by both theists and atheists for their illuminating treatment and popularisation of scientific topics. It may well be supposed by their readers that their comments on religion carry all the authority of their comments on the sciences. It needs to be shown, therefore, that their scientific statements do not carry the implications for religious belief that they may seem to do. In particular, the claims made for modern cosmology and for the theory of natural selection need to be examined with some care, to see where conflicts with belief in God may arise, and how they may be resolved.

The view I shall take is that, on most issues, there are no conflicts, and that the success of scientific investigation corroborates theism, rather than the reverse. For scientific investigation will only succeed if the universe is an intelligible and mathematically comprehensible unity, as theism supposes it to be. There are, however, particular conflicts that exist. One main one with which I shall be concerned is the claim of some cosmologists (including Carl Sagan and Stephen Hawking) that modern physics has somehow shown God to be at best superfluous and perhaps even to be an irrational construction. This claim can be countered, book for book, by other eminent scientists (Chris Isham, Paul Davies and John Polkinghorne come immediately to mind),[3] but it still has currency and popularity. In considering it, I shall take one superbly written book, *Creation Revisited*, by an Oxford chemist, Peter Atkins. I shall try to show that it actually contains many philosophical mistakes, and that when these are removed, it seems to argue more in favour of God than against God.

The other main conflict I shall be concerned with is between Darwin's theory of natural selection, as interpreted by neo-Darwinians like Richard Dawkins, and a theistic belief in creation. I should make it quite clear that I accept the view that life on earth, and indeed the whole present universe, evolved from earlier much simpler physical states. I accept a theory of evolution as one of the major insights of modern scientific understanding, and I think that it enriches traditional religious belief in God considerably. The dispute is about how evolution operates – whether by blind chance or by divine providence. I shall argue that the latter view is much the more reasonable, in view of the evidence. This will bring me into conflict with Richard Dawkins, in particular, and his brilliant restatement of 'the Darwinian worldview'. The Darwinian worldview is the view of evolutionary naturalism. It is old-style materialism writ large. It suffers the same disadvantages, plus some new ones of its own. Some of these will emerge in discussion of the views of Michael Ruse, founding editor of the journal *Biology and Philosophy*, who explicitly sets out to defend the truth of evolutionary naturalism. The argument of this book, then, is that a theistic interpretation of evolution and of the findings of the natural sciences is by far the most reasonable, that the specific arguments of Atkins, Darwin, Dawkins and Ruse on these topics are often fairly weak, and that it is the postulate of God, with its corollary of objective purpose and value, that can best provide an explanation for why the universe is as it is.

NOTES

¹ The phrase is from Michael Ruse in the Introduction to *Evolutionary Naturalism*.
² See Michael Polanyi, *Personal Knowledge*.
³ See C. J. Isham, R. J. Russell and N. Murphy (eds.) *Quantum Cosmology and the Laws of Nature*; Paul Davies, *The Mind of God*; John Polkinghorne, *Science and Creation*. For a masterly overview of the topic, see Ian Barbour, *Religion in an Age of Science*.

CHAPTER ONE

The Origin of the Universe

THE UNIVERSE: A BRUTE FACT?

According to the best available knowledge in modern physics, our planet earth, circling around the sun, a medium-sized star, was formed millions of years ago from stellar dust. The galaxy of stars to which our sun belongs is one of millions of galaxies, scattered throughout a space–time system which has itself expanded from a primeval burst of energy thousands of millions of years ago. This vast cosmos, awesome in beauty and in extent, began in a sudden blaze of energy from one singularity, a point of infinitely compressed gravitational force and density, which exploded in the primeval 'Big Bang'.

What existed then, at the beginning of this universe? Why did it begin, and why did it take the form it did? These are questions that have fascinated humans since the dawn of recorded thought. There are three main possible answers to these questions. One is that there is simply no explanation. The universe just came into existence by chance, for no reason, and that is that. Another is that it all happened by necessity. There was no alternative. A third is that the universe is created by God for a particular purpose. Those who do not believe that there is a God will have to take one of the first two answers. I shall try to show that neither of them is very plausible. The third, theistic, hypothesis of creation is by far the best explanation for the existence of the universe.

None of these answers requires the universe to have a beginning. Although most scientists accept that this universe, this space–time, did begin to exist, perhaps there were other phases of the expansion and contraction of space–time before it, so that maybe some time has always existed, without beginning. In that case, the same three general answers to the question of existence would still exist – either universes exist for no reason, by necessity, or for a purpose or set of purposes. Strictly speaking, therefore, the question of whether the time of this universe had a beginning is not relevant to the question of whether the whole universe is created, or whether it exists without a creator. The question is the same, whether time began or has always existed. The question is, what explains the existence of space and time, or is there no explanation?

Stephen Hawking is, untypically, being quite naive when he says, 'So long as the universe had a beginning, we could suppose it had a creator. But if the universe is really completely self-contained, having no boundary or edge, it would have neither beginning nor end: it would simply be. What place, then, for a creator?'[1] He presents a picture of the universe as leaving no room for God, who has been pushed out of the universe by the universal laws of nature. Hawking then suggests that maybe God has one last foothold on reality. God might be needed to start the whole process going. However, Hawking argues, if the universe had no beginning, God has been dislodged from His last hiding-place, and is rendered totally superfluous.

This picture that Hawking presents ignores completely the work of all major religious thinkers, who have agreed that what needs explaining is the nature of the universe as a whole, whether or not it had a beginning. They have, it is true, often used the word 'creation' to refer to the origin of this space–time. But, even then, they have always been clear that the distinction between creation, as God's act at the first moment of time, and the divine preservation of the universe in being at every moment of its existence, is a merely logical distinction. The crucial question remains: does the universe as a whole exist without having any reason or explanation, or because it has to be the way it is, or because it is brought into being and held in being at every moment by a supra-cosmic creator?

The first attempted answer, that there is no reason why the universe came into existence at all, is the answer given by philosophers like David Hume. We can never understand why the universe began, since there is no reason why it began. All explanations have to stop somewhere, with a statement of brute fact. If one has a very complicated state of affairs, one might well hope to explain it by resolving it into simpler elements. If one gets down to very simple elements, it may be held that one has reached the end of the line, that there is nothing more one can do in the way of explanation. The Big Bang is just about as simple a brute fact as one could wish for. So this looks like an end of the line of explanation. The Big Bang just happened, for no particular reason, and that is all there is to be said.[2]

Actually, the Big Bang is not as simple as it may seem to be. The universe began to expand in a very precisely ordered manner, in accordance with a set of basic mathematical constants and laws which govern its subsequent development into a universe of the sort we see today. There already existed a very complex array of quantum laws describing possible interactions of elementary particles, and the universe, according to one main theory, originated by the operation of fluctuations in a quantum field in accordance with those laws. It may or may not be possible to obtain a 'theory of everything', a very general law which covers all physical processes. But even if it is possible, that general law will include hundreds of subsidiary laws, governing the possible motions of elementary particles at various stages in the development of the cosmos. The laws will have to cover every possible eventuality, and be sensitive to millions of particular states of affairs, in all their complex particularity. It is disingenuous to say that such a remarkably rich and integrated set of laws is a 'simple fact'. And if it is said that these laws did not actually exist at the beginning of this universe's time, one has an equally complicated hypothesis that laws come into existence through time, yet all integrate amazingly to produce a coherent universe.

However simple the first moment of this universe's existence may be claimed to be, it is undeniable that the universe now comprises an incredibly rich and complicated set of entities. What is really in question is how such a very complex universe as this came into being. It is false to suggest that it is somehow less puzzling to have a long step-by-step building up of complexity than to have an instantaneous origin of complexity. If lots of bits of metal slowly assemble themselves on my doorstep by simple stages into an automobile engine, that is just as puzzling as the sudden appearance of an automobile engine on my doorstep. One has to account for the apparently well-integrated accumulation of gradual steps to complexity, just as much as one has to account for the existence of complexity itself. Explaining the complex by resolving it into a long succession of simple steps leaves the problem of explaining why all those simple steps accumulated in such an amazingly organised way. Saying that the very first step was rather simple is no help at all, when one at once has to add that it needed the addition of a huge number of co-ordinated simple steps to get the universe as we now have it. Of course, complicated things consist of lots of simpler things. The problem is how those things all go together to form one highly organised complicated thing. And that problem is made no easier by saying that they did so gradually, over time. If complexity needs explaining, it needs explaining, however long it took to get there!

Moreover, that there exists any matter/energy which is governed by any laws of physics (principles of regular succession) is, for the 'brute fact' believer, an extraordinary coincidence, since there might very well have been no matter at all, or there might have been no laws of physics, or there might have been something to which laws did not apply, or the laws might have ceased to exist soon after coming into being. That is, events might not have continued behaving in the regular and predictable ways that could be described by such laws. The fact that there are laws which continue to operate on matter/energy in regular and predictable ways is a rather surprising fact, which might well have been otherwise. The whole set-up is not quite as simple as it may seem.

When one considers all the elements involved in the Big Bang, it begins to look like an extremely complex event, and not a simple elementary fact at all. So it still seems to stand in need of explanation. To say that such a very complex and well-ordered universe comes into being without any cause or reason is equivalent to throwing one's hands up in the air and just saying that anything at all might happen, that it is hardly worth bothering to look for reasons at all. And that is the death of science.

The whole of science proceeds on the assumption that a reason can be found for why things are as they are, that it is the end of science if one finds an 'uncaused' event, or one for which there is no reason at all. The Bohr/Heisenberg thesis that some quantum events occur without a cause seems at first to count against this claim. But it turns out that what is needed is a greater subtlety in thinking about what a reason is. Quantum uncertainties are governed by statistical laws. It is not the case that absolutely anything can happen at any moment, at the quantum level. Electrons cannot wholly disappear without having any effect on the rest of physical reality. There are rigorous laws of probability which describe quantum events.

At any one moment, there is a definite and finite set of possible futures for elementary particles. What Heisenberg claims is that not every event, at the quantum level, is *sufficiently* caused. There are many causes of quantum events, many factors which provide the conditions for their occurrence and which influence their occurrence. There is a certain indeterminacy about quantum processes. Yet the process as a whole is far from random. The laws of probability work so that most indeterminate and therefore unpredictable events cancel out at the macro-molecular level, and leave the highly predictable laws of mechanics intact. In dynamic systems far from equilibrium, however, small unpredictable changes can give rise to major changes of direction in the way things go.[3] This means that there is a certain 'openness' to the future in the natural order, that things are not wholly predetermined in every detail.

At a relatively late stage of the evolutionary process, moral freedom comes into being. Moral freedom is, broadly speaking, the possession of the ability to choose either good or evil. If one is to be blamed for choosing evil, one must believe that one could have chosen good, but failed to do so. That is, in the very same physical situation, one could have acted otherwise. That is only possible if physical events are not completely determined, if there is an openness in natural processes which permits real alternatives to exist.

Thus, there is a very good reason why not all physical events should be sufficiently determined by their physical antecedents (to be 'sufficiently determined' means that, given the antecedents plus some general laws, nothing else could happen except what does happen). The reason is that there can only be an open future if there is a degree of indeterminism. There can only be the sort of freedom that is morally important if there is an open future, at least sometimes. So indeterminism is a necessary condition of the later development of morally important freedom in rational beings.

I should perhaps make it clear that I am not making the reality of freedom depend upon quantum indeterminism. What freedom requires is simply that not all physical events are sufficiently determined. This could be for many other reasons than quantum indeterminacy. It could, for instance, be because many laws of physics are more like constraining limits than like sufficiently determining rules. It could be because not all physical events occur in accordance with the sort of processes of measurable and universal regularity which laws of physics ideally describe. All I am saying is that a good reason can be found for the existence of quantum indeterminacy, so that such indeterminacy does not contravene the scientific postulate that a reason can be found for why things are as they are.

This account entails that no reason can be given, for instance, for why a particular radium atom disintegrates at a particular time, rather than at some other time – that is one instance of what indeterminacy means. Yet a reason can be given why such indeterminate processes exist, and are limited by a precise and particular set of probabilities. There is a reason why things are as they are, though this actually precludes there being a reason for every specific event. That is the nature of a probabilistic universe, which seems to be the sort of universe we inhabit. It is very different from a wholly random universe, or one for which there is no reason at all.

The other thing worth noting at this stage is that the sort of reason just given – that indeterminacy is a condition of the emergence of moral freedom – is a *teleological reason*. It gives the reason for an occurrence, not in some previous state plus an impersonal, value-free law, but in some future value (moral freedom) which can be appropriately actualised by some present process (indeterminacy). Teleological reasons are very familiar in human life. A good reason for my writing this book, for instance, is my desire to have the ideas in it critically discussed and thus, with luck, contribute to the sum of human understanding. I explain the present activity in terms of a future value or desired goal.

Many physicists might protest that teleological reasons

– citing desirable values and the conditions of their actualisation – have no place in science. To that I would make two responses. First, scientists often, and rightly, press their questions to the ultimate degree, to the question, 'Why are the ultimate laws of nature as they are, and why is the initial state of nature as it is?' At that point, the obvious sort of reason to offer is precisely a teleological reason, which would state how the initial state and the laws together are well formed to actualise states of value. As John Leslie says, the ultimate reason why things are as they are is likely to be: because they actualise great and distinctive values.[4]

Second, scientists often do appeal to teleological reasons, to a sense of beauty and elegance, in choosing ultimate theories. Thereby they seem to commit themselves to saying that the universe exists because it is beautiful, and that might be an ultimate reason for its existence. Steven Weinberg writes, 'There is a beauty in these laws that mirrors something that is built into the structure of the universe at a very deep level.'[5]

I am not trying to construct a direct argument to God from such considerations. Neither John Leslie nor Steven Weinberg believes in God – though that is largely, I think, because they cannot connect the idea of the ultimate value of beauty with what they (wrongly) see as the very anthropomorphic and rather sentimental God of religion. My argument is that science is based on the postulate that one should always seek reasons for why things are as they are. If, every now and again, things just happened for no reason at all, not even for probabilistic reasons, science would come to an end. If I ask, 'Why does water boil as it is heated?' I do not expect to be told, 'There is no reason at all. It just does.' Not much physics would get started with that attitude, and not many examinations in physics would be passed by candidates who gave that sort of answer. So it does seem a little odd that a physicist may work all the way back to the Big Bang, assuming there is a reason for everything, and then say, 'Well, at this point, there is no reason at all. It just happened.' It seems odd to think that there is a reason for everything, except for that most important item of all – that is, the existence of *everything*, the universe itself. The physicist might actually expect a reason to exist. Of course, the physicist may just have to be disappointed. It may be, I have suggested, that an ultimate reason would have to lie beyond the strict limits of physics itself, in some sort of teleological reason. But the 'no reason' hypothesis should be a hypothesis of last resort. It should only be accepted when everything has been tried, and found wanting. This is the least tempting of the three ultimate hypotheses.

THE UNIVERSE: AN UNDERSTANDING OF EVERYTHING?

So some physicists, like Steven Weinberg, unhappy with a resort to 'mere chance', which could never be rationally understood, postulate a second main answer to the question about the origin of the universe. They suppose that it came to exist, not by chance, but by necessity, and so it can, in principle, be completely understood.[6] One variant of such a view is that there is only one logically consistent set of quantum laws which, operating over some form of primeval energy, inevitably gives rise to a universe like this sooner or later.[7] This hypothesis could never be proved, since there may be infinitely many forms of the consistent organisation of matter of which we have no conception. It is virtually impossible to prove the negative proposition that no system other than the ones we can think of could possibly exist. To say that the existence of this universe is necessary is to say that no other universe could possibly exist. But how could one know that, without knowing absolutely everything? Even the most confident cosmologists might suspect that there is something they do not know. So it does not look as though the necessity of this universe can be established.

It is true that mathematical laws, if they exist, must exist in a quasi-Platonic realm of pure thought. They have to exist, and cannot be other than they are. This may seem to provide the basis for necessity. But what about the existence of the matter/energy, the basic stuff to which the laws apply? For Plato, matter was just a 'surd' element, always existing, a sort of unformed chaos, which could be put into shape by the application of laws (by whom?).[8] Explanation is still incomplete, if we end up with brute matter, which does not have to be the way it is. Things seem to be no better than when we started.

The physical cosmos does not seem to be necessary. We can seemingly think of many alternatives to it. There might, for instance, be an inverse cube law instead of an inverse square law, and then things would be very different, but they might still exist. We can see how mathematics can be necessary, but it is a highly dubious assertion that there is only one consistent set of equations which could govern possible physical realities. We cannot bridge the gap between mathematical necessity and physical contingency. How could a temporal and apparently contingent universe come into being by quasi-mathematical necessity?

One attempted answer to this mystery is put forward in Peter Atkins' book *Creation Revisited*. It begins with the assertion that 'there is nothing that cannot be understood'.[9] We can even understand why the universe began. This is a remarkably bold statement of faith. It goes well beyond all available evidence, since at present there are millions of things we do not understand, including the fundamentals of quantum physics.[10] It is, I suppose, partly based on evidence, on the success of the natural sciences. But it is mostly based on a faith in the power of the human mind to understand things correctly, and in the rational structure of reality, which is inherently knowable, mostly by mathematical means.

While this faith goes beyond the evidence, it is not irrational or blind, even though it may have an element of scientific wish-fulfilment about it. One might perhaps call it a fundamental postulate, which underlies and motivates the investigation of the natural universe, and gives hope for the success of scientific study. While it may be misplaced, such faith is a foundation of rational activity, and one might well wish to commend it on grounds of its practical utility. It should not be thought of as merely useful, however, as though it only pandered to psychological needs or desires. On the contrary, it actually defines the fundamental values to which our needs and desires should, in the sciences and in the human search for truth generally, be conformed. This postulate of faith is actually a fundamental value-commitment, and the values in question are the traditional triad of truth, beauty and goodness. Faith in the comprehensibility of the universe is in fact faith in the ultimate truth, beauty and goodness of reality, in the virtue of pursuing them and in the certain hope of eventually finding them.

It is faith in truth, because it postulates that the human mind can formulate and understand an objective truth about the way the world is, whether or not it is to our liking. It is faith in beauty, because the criteria often used in the intellectual search are those of the simplicity, elegance and beauty of the basic laws of being.[11] It is faith in goodness, because it presupposes that the universe itself is 'friendly' to our investigations, allowing itself to be understood and in that way fulfilling the deepest potentialities of our intellectual natures.

It is clear that Dr Atkins begins with a faith of just the same nature as religious faith, a fundamental postulate of the intelligibility, beauty and (mathematical) harmony of the universe, and of the possibility of human fulfilment in understanding its own ultimate environment. Monotheists will immediately recognise this faith as their own, and may even claim without self-deception that faith in science, in the rational structure of nature, has historically been strongly motivated by faith in a wise God who would be expected to provide such a structure.[12]

I suppose, however, that most theists would demur at saying that everything can be understood by human minds. Reality, a theist might say, is intrinsically comprehensible, but is completely comprehended only by the supreme intellect of God. But maybe *total* understanding of *everything* is a bit much to ask of a tiny human mind. Theists do not, or should not, put a barrier up in front of scientists, and say, 'Do not try to understand this; it is forbidden knowledge.' On the contrary, they should say, 'God has created you to understand and revere creation; therefore seek truth as vigorously as you can.' Yet theists might also want to say that there is not just one sort of understanding. As well as the cool, dispassionate analyses of science, there is the passionate, self-involved contemplation of art, the search for moral truth, and the search for the ultimate reality, the creator. In such pursuits, there is a place for mystery, for that which is beyond intellectual analysis, and yet surpasses the finite, abstractive and discursive intellect by an intelligibility that goes infinitely beyond its powers.

THE ABSTRACT WORLD OF PHYSICAL SCIENCE: THE FALLACY OF MISPLACED CONCRETENESS

It is important to see the limits of the human mind, as well as its possibilities. One limit is that the mind works by abstraction and generalisation. Reality as it is, and as it is immediately experienced, is a presentation of particular, complex, uniquely contextualised facts. Each experienced element is unique in its placing within a stream of experience. It is particularised by a perceiving response which carries with it a whole battery of past learned responses and dispositions to future action. It is interwoven experientially with an array of other elements which affect its character and the meaning it has for the percipient. The immediate particularity of experience is incommunicable by language, and it is in the arts that it may be indirectly conveyed or evoked.

Modern science began with the insight that one could abstract from this array of unique particular elements, and construct quite general formulae, expressive of significant relationships between groups of experienced elements. Thus a simple mathematical expression such as '$E = mc^2$' (Energy equals mass times the square of the speed of light) can express an abstracted relation which holds for all cases that fall under a general description such as 'E' and 'm' (i.e. which have mass and energy). Such abstraction is essential to human understanding, and it has opened up comprehension of natural processes in an amazing way. But it should not be forgotten that it is an abstraction. It will hold true only insofar as one can distinguish and isolate precisely specifiable elements within experience, and state general relations between those elements.

The first scientific task is to isolate such elements as can be quantified and related in a useful way – in the case of Newtonian physics, the elements of mass, position and velocity. Nature is kind to us, in that it does contain elements related to one another in constant and mathematically quantifiable relations. This enables us to achieve great predictive power and control over physical processes. When one states such relations, in the fundamental equations of physics, say, one is abstracting elements of a certain sort, for a specific purpose (of 'seeing how things work, and so how to work them'). It is possible, particularly in highly sophisticated mathematical constructions like quantum field theory, to construct totally abstract mathematical schemes, which have high predictive value for highly specified and tightly controlled experimental situations. The mathematical physicist and philosopher A. N. Whitehead pointed out, however, that this can easily give rise to what he called the 'fallacy of misplaced concreteness'.[13] That is, one is so impressed by the mathematical elegance and predictive power of one's own construction that one comes to see *it* as the true reality, while the phenomenal experience from which the scheme began is relegated to the realm of mere subjective illusion. This is the ultimate irony of some modern science, that it begins by trying to explain and understand the rich, particular, concrete world as experienced by humans, and ends by seeing that phenomenal world as an illusion. The true reality becomes a realm of such abstract entities as Hamiltonian vector fields in multi-dimensional phase space, which we can scarcely imagine, let alone experience. The fallacy is to take the abstract for the concrete, to see our construct as the only reality, and to see experienced reality as a product of confused perception.

This fallacy has snared philosophers from Plato to Leibniz and beyond, and it still snares many major physicists. It is perhaps important to emphasise that it is a philosophical theory. It is not an assured result of science, but a particular interpretation of the data that science provides. It is indeed a remarkable fact that the mind can construct, out of the seemingly analytic processes of pure mathematics, a model of the universe that can predict certain classes of future events with fantastic accuracy. I am myself not averse to interpreting this, as the Oxford mathematician Roger Penrose does, as a confrontation of the mind with a Platonic world of pure mathematical forms.[14] For a theist, this is indeed quite a natural interpretation, for where else would the mathematical forms be but in the mind of God? And what other than the power of a creator could ensure that physical processes at least partly express a subset of such forms? What goes wrong is that the Platonic world – the world of 'ideas in the mind of God' – is seen as the truly real world, and the world of sense-experience, of personal relationships, of beer and skittles and suffering and love, is demoted to a shadowy half-reality. An alternative view is to say that the Platonic world provides the intelligible structures and relationships of this and of infinitely many other possible universes. But this actual universe is a universe of particularity and consciously apprehended uniqueness. Many of its fundamental elements are governed by a subset of Platonic possibilities. Those mathematically picturable relations are the skeleton of the universe. Or perhaps, to change the metaphor, one might say that mathematical physics provides a map of the universe, according to a particular projection. But skeletons are not living forms, and maps are not landscapes in which one can breathe and walk. A landscape is of quite a different nature from a map, and it contains many different sorts of elements which maps are unable to include, because of their nature as abstractions.

THE RELATION OF MATHEMATICS TO THE WORLD: THREE FALLACIES

Atkins espouses the fallacy of misplaced concreteness in the grand manner. He entitles his version of it 'the hypothesis of strong deep structuralism',[15] which is the thesis that 'physical reality is mathematics and mathematics is physical reality'. This identification of opposites is just about as extreme as one could get. The mathematical realm is the realm of necessary, timeless, abstract and precise universals. The physical realm is the realm of contingent, temporal, concrete and fuzzy particulars. Most philosophers find one of the intractable mysteries of their art is to state in a reasonably correct way the relationship between these realms. The mystery can indeed be dissolved by saying that they are the same thing; but it would require an invincible argument to drive one to such an obviously false conclusion.

The argument Atkins actually offers appears on page 113 of his book. It may be helpful to take it step by step.

He begins by saying that: 'Aspects of the universe are summarized by mathematical formulas.' There is nothing wrong with that.

He continues: 'Formulas are generalized statements about the relations between quantities.' At this point a hesitation arises. If one takes the famous formula '$E = mc^2$', would one say that it is *about* a relation between quantities? It is certainly about the relation between energy and mass. It states that the amount of energy in a physical system is equivalent to the mass times a constant, equal to the square of the speed of light. One could use the word 'quantity' for energy and mass, as long as one remembers that one is speaking of a certain measurable quantity of two different properties, mass and energy. It would be clearer, however, to say that formulae are about the relations between measured *properties*. They are not about the variables 'E' and 'm', which in themselves have no particular numerical values. And they are not about any two particular numbers which might be substituted for those variables in particular cases. They are about the relation between two different properties, which have a constant ratio to one another.

The argument proceeds: 'Those quantities are expressed numerically.' There is no problem here, as long as one remembers that it is the measured quantity of a specific property that is being expressed numerically, not just a quantity in the abstract.

Finally, and with an unashamedly fallacious jump, the argument produces the conclusion: 'Hence, formulas are statements about the relations between numbers.' The 'hence' is quite misplaced. This is a huge *non sequitur*. All one can validly infer from the three premises, properly stated, is that mathematical formulae are about relations between properties whose measured value is expressed numerically.

What has gone wrong is that a small equivocation over the term 'quantity' has led to an invalid move from saying that formulae are about numerically expressible relations between measurable properties, to saying that formulae are about the relations between the numbers in which such relations are expressed. Formulae of physics are *expressed in* numbers, but they are not *about* numbers. They are about relations between physical properties.

The mistake is like the similar mistake, often made, of arguing that, since language is about God, and language is a human construct, therefore God is a human construct. However, what language or mathematics *is* (a human construct), and what it is *about* (God, or the physical world), are two different things. It is only by confusing them that Atkins can hold that mathematics and physical reality are identical.

His motive in this is presumably to eliminate the intractable reality of the physical, leaving only the mathematical, which itself then resolves, by the application of set theory, into 'assemblages of absolutely nothing',[16] thus leaving nothing for a creator to create. But if the reduction of many things to nothing can only be accomplished by a succession of fallacies, one may think that there is still quite a lot for a creator to do, after all. Three of these fallacies have just been exposed, the fallacy of misplaced concreteness, a fallacy of equivocation over the term 'quantity', and what may be called the 'intentional fallacy', of identifying what an expression is *about* (its 'intention' or reference) with what it *is* (a verbal or mathematical construct). Four further fallacies will soon appear, as we explore Atkins' remarkable claim that everything is actually nothing.

THE LIMITS OF HUMAN UNDERSTANDING

If one can resist the claim that the mathematical realm is the only truly real realm, one can see that scientific comprehension, at least in physics, is a comprehension of some of the basic relational elements of the physical structure of the universe – those which are related in quantifiable and constant ways. What it gives is a generalised and abstract picture of the universe. What it does not, or should not, claim is that it is a complete description, in the sense of including every existent element, or that it describes what the universe really is, whereas everything else is an illusion. Abstraction is a wonderful property of the human mind, which has given rise both to language and to modern science. It should be balanced, however, by that attention to particularity and concreteness which art and, at its best, religion encourage. If it is not, it becomes something of a handicap in the search for the full truth of things, and in that sense it can be a limitation of the human mind.

A further important human limitation is that the intellect works discursively. That is, it is incapable of intuiting things in one all-embracing experience. It has to consider things one after the other, making connections by inference and extrapolation, and moving from one element to another in succession. A fully comprehensive intellect, like that of God, will understand all things in one intuitive, non-discursive, act. God does not need to infer or extrapolate, since God knows everything in its full particularity by immediate apprehension. That sort of knowledge is not possible for humans. So this is another respect in which human minds can never understand absolutely everything, in all its fullness, as it really is.

Finally, it should be obvious that there may be many universes, that is, finite space–times, and forms of existence other than those in this space–time. If God is infinite, there is presumably an infinite number of things to be understood, before everything is understood. There is absolutely no way in which we could have knowledge of other universes (for by definition they would have no spatial or temporal relation to us, which rules out all forms of knowledge), and there is no way in which a finite mind can encompass an infinite set of data (unless they could be known to be infinite repetitions of a finite class of data, which is not the case). So it seems, after all, that if everything can be understood, only a God could understand it, and so Atkins is committed to theism. In fact, I am rather puzzled by the fact that he does not seem to realise it.

NOTES

[1] Stephen Hawking, *A Brief History of Time*, p. 141.

[2] Not many physicists are happy with such an idea. Among philosophers, David Hume is its best-known exponent, in *A Treatise of Human Nature*, esp. Book 1, part 3, section 3 (first published 1738). Hume's account of causality is rejected by most philosophers.

[3] See the work of Ilya Prigogine, for example in I. Prigogine and I. Stengers, *Order out of Chaos*.

[4] In *Value and Existence*, the physicist John Leslie writes that the world exists 'as a result of an ethical need' (p. 160).

[5] Steven Weinberg, *Dreams of a Final Theory*, p. 194.

[6] 'We would prefer a greater sense of logical inevitability': Steven Weinberg, *The First Three Minutes*, p. 76.

[7] Among others, such a view has been suggested by C. M. Patton and J. A. Wheeler in 'Is Physics Legislated by Cosmogony?'.They admit their lack of success in producing a complete theory of this sort.

[8] Plato, *Timaeus*, pp. 48ff., pp. 65ff.

[9] Peter Atkins, *Creation Revisited*, p. 3.

[10] 'I think I can safely say that nobody understands quantum mechanics': Richard Feynman, *The Character of Physical Law*, p. 27.

[11] 'Dirac is unabashed in his claim that it was his *keen sense of beauty* that enabled him to divine his equation for the electron': R. Penrose, *The Emperor's New Mind*, p. 545.

[12] This case is argued strongly by Stanley Jaki, in *Science and Creation*, and in his Gifford lectures, *The Road of Science and the Ways to God*.

[13] A. N. Whitehead, *Science and the Modern World*, pp. 64ff.

[14] Penrose, *The Emperor's New Mind*, ch. 10.

[15] Atkins, *Creation Revisited*, p. 109.

[16] Ibid., p. 115.

CHAPTER TWO

Something for Nothing:
A Dubious Deal

SOMETHING FROM NOTHING: FOUR USEFUL LOGICAL TRICKS

Perhaps the reason for Atkins' failure to understand the implications of his argument lies in his second postulate of faith, that 'everything is extraordinarily simple'.[1] This postulate has that air of paradox typical of religious propositions. To put it bluntly, Atkins would not bother to make the statement unless it was, in a sense, obviously false, and so slightly shocking. Like some profound religious utterances, however, its real function is to startle us out of our everyday acceptance of things, and propose a new vision of an underlying reality which stirs wonder and awe. I suppose that ultimate simplicity is nothingness; it would be hard to get simpler than that. And indeed Atkins proposes that 'the seemingly something is actually elegantly reorganized nothing, and . . . the net content of the universe is . . . nothing.'[2]

The word 'nothing' has long been a favourite of those who wish to distil apparent wisdom out of empty verbal juggling, and I fear this is no exception. To say that nothing exists is to say that it is not the case that there is anything that falls under any description. Yet the verbal form, 'nothing exists', traps the unwary into thinking that there is at least one thing that exists and falls under some description, namely 'nothing'. This is a good example of the 'reification fallacy', which is the error of thinking that all nouns must refer to something, therefore 'nothing' must refer to a peculiar sort of something. By committing this fallacy, one performs the first logical trick needed to produce the conclusion that this very rich and complex universe is actually 'elegantly reorganized nothing'. This is the claim that when one says that nothing exists, one is saying that at least one thing, nothing, exists.

The second trick is to say that, since this nothing exists, it can have various properties, as long as these properties cancel each other out, as the numbers '1' and '-1' do, or as positive and negative electric charges do, or as matter and antimatter do.[3] So 'nothing' turns out to consist of an equilibrium of an infinite number of equal and opposite forces, charges and so on.

The third trick is to suppose that, since this huge number of forces in equilibrium exists, fluctuations may arise in it, whereby tiny disequilibria appear from time to time. In that way, it is suggested by Atkins, something (tiny positive or negative forces) may arise, by chance, out of absolutely nothing. But in order to resist falling victim to this trick, one must hold fast to the thought that where quantum fields exist, there is hardly 'absolutely nothing'.

The fourth trick is to suggest that one of these fluctuations may bring four-dimensional space–time into being, so that any need for a creator disappears, and one can see that the universe comes from nothing and consists of nothing. There is, quite simply, nothing to create and therefore no need for anyone to create it. (All this is summarised on page 149 of Atkins' book.) The trick here is the inference from 'universe x comes from nothing' (i.e. has no cause), to 'the universe consists of nothing' (i.e. is not there). It may be possible to think of a universe springing into existence out of nothing at all. But once it is there, it is most certainly something.

The accumulation of these four logical tricks is what enables Atkins to say that 'the universe is nothing'. However, if it is not the case that there is anything which falls under any description, then it cannot be the case that there exists an infinite set of perfectly balanced positive and negative forces. It is a mistake to say that even possibilities of any sort exist, if there is absolutely nothing.

I am inclined to say that possibilities do exist. Even if no actual universe existed, its possibility would exist, together with the possibilities of every other possible universe, all comprising an infinite set of possibilities. We are back to the Platonic world of pure forms, pure possibilities. But how can mere possibilities exist? One must be logically ruthless, and say that either there are really no possibilities or that they exist in something actual. In that case, since possibilities will always, eternally, exist, there never is absolutely nothing. There is always something, and something that can contain in itself every possibility. A definition of a 'necessarily existing being' is that it is a being which exists in every possible logical world (where 'a world' is taken to cover absolutely everything that actually exists). If, wherever anything is possible, there exists an actual being which contains that possibility, and if that actual being is the same in all possible worlds, then by definition that actual being is a necessarily existing being.

Now all possibilities, as possibilities, are necessarily existent. That is, every possible world exists in every other possible world, precisely as a possible, but usually non-actual, world. Therefore the actual being that contains each possible world is one and the same being that contains all possible worlds. So there is one and only one actual and necessarily existing being. It will itself exist of necessity, and it will necessarily contain all possibilities in itself.

In this way, one might hold that the second hypothesis that seeks to explain why the universe is the way it is, the 'necessity hypothesis', does lead to the postulation of an ultimate being which exists necessarily and which has a necessary nature. But that being is not the physical universe, which seems both dependent and contingent. And it is not some realm of pure mathematics, which is in itself a mere abstraction. Theists call it the mind of God, and the doctrine of creation is that this universe comes into being out of nothing except the mind of God. In this way, the second hypothesis naturally leads on to the third hypothesis, the theistic hypothesis, that there is a being which exists of necessity, but which creates this universe by a free act of will. God bridges the gap between the necessity of the conceptual realm and the contingency of the physical realm. For God exists by necessity, with an essential nature which is necessarily what it is – including the total set of possibilities that exist in the mind of God. Yet God's necessary nature can be expressed in many contingent ways, any one of which might properly manifest the love, creativity and wisdom that comprise the essential nature of God.

The idea that a being can be necessary in some respects and contingent in others may sound surprising. But it can easily be shown to be perfectly coherent. The divine existence, goodness, omnipotence and omniscience are all necessary properties, and God knows all possible worlds, as they necessarily exist in the divine mind. If one had perfect understanding (as God has), one could explain the existence of God by showing how it, and all the essential properties of the divine being, are necessary, and could not fail to exist or be other than they are. In this way, Weinberg's desire for a 'greater logical inevitability in explanation' (see page 33, note 6) can be fulfilled.

God also has many contingent properties, and some of these are entailed by God's necessary properties. For instance, the necessary property of being omnipotent entails that God is able to do an unlimited number of things. God can create a world in which I die young or a world in which I live for many years. If God is omnipotent, God can do either of these things, but of course not even

God can do both. It follows that a necessarily omnipotent God must be able to act in contingent ways, ways to which there are real and coherent alternatives.

In general, the vast majority of the acts of God concerning a created universe will be contingent, as will be the creation of any particular universe. If there is a God, one could (with perfect knowledge) explain the existence of a particular universe by showing that God chose to create it, for a good reason – for the sake of the goodness it actualises, and which God and many creatures can apprehend and enjoy. This is a teleological explanation, which is the only sort of explanation a truly contingent universe can have. In this way, the requirements of both necessity and rational choice can be seen to be met by a value-actualising universe which is created by a necessarily existing God.

Atkins tries to get a universe out of nothing. What he refers to as 'nothing', however, is actually a rich realm of possibilities, from which an actual universe somehow arises. My response has been to say that the postulate of such a realm of possibilities is not absurd, but that it is best conceived as located in the mind of God. The divine mind does not come into being, since it always exists, as the conceiver of all possibilities. If it has come into being, it would once have been merely possible. But then, *ex hypothesi*, it would already have to have been conceived in the mind of God, which cannot therefore have been merely possible. So the mind of God can never come into being. Like numbers, it must always exist, and it can be the source of any actual universe. The mind of God is not 'nothing' (though one may well say that it is not a – finite – thing). It is a necessarily existing source of all actuality, which actualises a subset of possibilities by a contingent act of will. That is where Atkins' argument ought to lead. It does not eliminate God at all. On the contrary, it provides a rather elegant way of getting at an important part of what is meant by the word 'God'.

How nothing can fluctuate: an incredible proposal

Atkins seeks to eliminate any talk of an act of will on God's part by supposing that physical realities can arise out of purely conceptual or mathematical realities by some sort of natural necessity. He supposes that fluctuations may arise in nothing, and sooner or later produce a physical universe. This proposal is based on various ideas which have been developed in quantum cosmology. As soon as one examines these ideas, however, it becomes clear that there is something fluctuating, after all. A 'quantum fluctuation' is a non-determined change in such properties as position, momentum and energy, which occurs in the microworld of subatomic particles. Such changes are governed by statistical laws, but are not absolutely determined by preceding causes. Quantum cosmologists suppose that the whole universe can be treated as subject to quantum effects, especially at a point very near the Big Bang, say before the Planck time, 10^{-43} seconds after the Big Bang. At that time, the universe would have been small enough (10^{-33} cm across, the Planck length) to be subject to quantum effects. So in that domain one might expect non-determined energy fluctuations.

It is also possible to hold that the universe as a whole has zero net energy. For gravitational energy is negative, while rest mass and kinetic energy are positive. If these energies balance out, one has zero net energy. If, in this state, there are quantum fluctuations, then one might say that the universe could come into existence out of 'nothing', in the sense that it has come in a 'chance' way out of what physicists call a 'vacuum' (quantum fields in their ground or lowest energy state). However, this is all very far from 'absolute nothing'. There has to be a background space–time, to allow fluctuations to occur. There have to be quantum fields with very definite properties of energy, mass and so on. The fact that these properties 'balance out' to give a zero sum is rather like the fact that if an accountant's books balance exactly, they end with a zero. An enormous amount of activity has actually taken place in the meanwhile, and an accountant's zero is totally different from the zero of a person who really has no money at all. Similarly, the cosmologist's 'zero energy state', with its perfect balance of gravitational field and kinetic energy and rest mass is as far as one can get from absolute non-existence.[4]

Finally, one has to have in place the probabilistic laws governing quantum fluctuations. While they are non-deterministic, such laws do assign probabilities to states of affairs with precision. Naturally, if they are thought of as really existent, even before a physical universe exists, they will guarantee that a universe will 'fluctuate' into existence with a finite probability, and therefore within a finite time. But a law that ensures the existence of a universe sooner or later can hardly be said to make the existence of that universe a matter of 'chance'. There is a weighted probability towards the existence of the universe built into the quantum laws already.

On the quantum fluctuation hypothesis, the universe will only come into being if there exists an exactly balanced array of fundamental forces, an exactly specified probability of particular fluctuations occurring in this array, and an existent space–time in which fluctuations can occur. This is a very complex and finely tuned 'nothing'! Nor is it a purely conceptual or mathematical state, which itself exists by necessity. Many other arrays and sets of laws are possible. So this universe looks highly contingent after all, and a creator God might well choose to create a partly probabilistic universe by choosing just such a mode of origin for it.

An alternative cosmology is the Hartle/Hawking model, which does not assume a background space–time in which the universe arises. The details of that model depend upon a theory of quantum gravity, which is as yet not much more than a tentative proposal. But in it 'external time', as a dynamic reality carrying a sense of 'passage' and of development, is replaced by a concept of 'internal time', identified with such things as the curvature or temperature of a particular three-dimensional space. According to the Hartle/Hawking model, a quantum mechanical state function 'Y (c,f)' is formed, giving the probability of finding a particular curvature, 'c', or matter field, 'f'. Our ordinary sense of time is then seen as one possible way of 'sewing together' three-dimensional spaces. But time itself is signified by a complex number (part of which involves an imaginary number, such as the square root of a negative number), and it becomes an internal property of a set of three-spaces.[5] I do not think this can any longer rightly be called 'time' at all, in any sense we can recognise it. What has happened is that the phenomenological reality of time has been transformed into a mathematical variable, and then treated as a pure abstraction, which, far from giving the 'true reality' of time, has less and less relation to the real time one started from.

Nevertheless, according to the model, time itself seems to arise (though obviously not in a temporal sense) from a more spatialised quantum gravity domain. This is still not an 'origin from nothing'. It requires pre-existent Hilbert spaces, quantum operators, Hamiltonians, imaginary numbers and other abstract mathematical entities which are, if anything, more mysterious than the existence of the space–time universe itself. The conceptual problems of such a model are enormous.

With this model, one is clearly not talking any longer about 'fluctuations' in an ordinary sense, entailing change in time. If external time does not exist, no state can follow any other state, and it looks as though one must have a totally static realm, in which fluctuations can have no place. Atkins speaks of 'the dust from which spacetime is to be built'.[6] But we must put out of our minds any idea of quasi-physical dust, scattered randomly but assembling itself into various structures from time to time. As he notes, what is meant is 'a Borel set of points not yet assembled into a manifold of any particular dimensionality'.[7] Strictly, the word 'yet' should be removed from this sentence, as it connotes a temporal dimension. Then one sees that one is dealing with a purely mathematical idea of some complexity, which does not in any sense *precede* a more definite structure of physical entities, and which can have no *causal* relation to any such structure. There is a great danger here of confusing the mathematical (the abstract, timeless and relational) with the physical (the causal and temporal). This is again the fallacy of misplaced concreteness, this time taking the form of the fallacy of regarding numbers as physical entities. It might well be called the 'picturing fallacy', because it treats purely mathematical relationships or abstract concepts as though they were literally pictures of physical entities. In doing that, it misconceives what abstract mathematical thought is. A 'dimension' in mathematics is simply a co-ordinate, and it does not necessarily have any physical correlate. A 'dimension' in the physical realm is an extension at right angles to another extension. If time is regarded as a dimension, this is quite a different, analogical sense of 'dimension', and already the usefulness of a mathematical formalism may threaten to produce misleading metaphors (of time as a sort of successionless state, a spatialised time), when translated into concrete imagery. When Atkins says, 'Think of the primordial dust as swirling',[8] the metaphor has replaced the reality, and a fallacy has been taken to be a profound truth.

Can we imagine a timeless fluctuation at all? It will certainly not be a 'swirling', a 'gathering' or a 'stumbling' – all terms used by Atkins. What will it be? What one can do (what mathematicians do) is to think of an array of three-spaces, representable only by sophisticated mathematical terms involving complex numbers and quantum probabilistic wave-functions, set out alongside one another. There is no movement and no process. One can speak of probabilities, in terms of the clustering of solutions around certain values, and of tendencies to converge or form patterns. But these are terms of art, and they all refer to, and must strictly be translated into, patterns of timeless relations, which have a complexity and order so astonishing that it leads many mathematicians to speak of the Platonic realm as having its own kind of reality.

On the Hartle/Hawking model, classical space–time becomes a domain of the universe in which many three-spaces can be smoothly joined together, where the state function 'Y' is maximal. There are many other domains where time as we know it does not exist. But does this mean that, as Hawking claims, 'The universe would be completely self-contained and not affected by anything outside itself. It would neither be created nor destroyed'?[9] Absolutely not. The question remains: what, if anything, accounts for this amazingly complex set of three-spaces and all their interrelationships, of which this space–time now becomes just a part? It is only because Hawking assumes that 'creation' means 'a beginning of time' that he can say the universe as a whole is not created. It is only because he does not ask why the quantum laws are as they are that he can say that the universe is not affected by anything outside its own parameters. We cannot possibly imagine how many other, quite different, sets of quantum laws might be possible in addition to the ones we can conceive. And even with the set of laws we think we can conceive, the actual existence of this space–time seems, far from being certain sooner or later, to be immensely improbable. Roger Penrose suggests that the probability of a universe like the one in which we exist, out of the array in 'phase space' of possible universes, is 1 in 10^{123}, a number too big to be written down in full even if every proton in the entire universe were used to write a digit on![10] The physical existence of this universe, even on highly disputable quantum gravity theories such as those of Hawking, is due either to extraordinary chance or to a choice from possible mathematical structures of extraordinary precision. Quasi-mathematical necessity cannot of itself give rise to an actual universe. The hypothesis of chance takes one back to the unpalatable first hypothesis for the origin of the universe. Once again, the argument actually seems to point in the direction of intelligent choice rather than blind necessity, to a designing mind rather than to a quasi-mathematical reality which somehow gives itself physical embodiment. The necessity hypothesis once more needs supplementing with the theistic hypothesis, that the necessarily existing being, God, freely creates a universe for a purpose.

HOW SOME POSSIBILITIES ARE IMPOSSIBLE

What Atkins is perhaps suggesting is that, in infinite time, even this extraordinarily improbable universe would come into being sooner or later, so that it is a matter of purposeless chance which nevertheless gives rise to this universe by necessity. But 'before' the existence of the space–time of this universe, we do not have infinite time, or any time at all! Is the suggestion, then, that every possibility that can be actualised, is actualised? This seems a highly profligate view of universes, and it does not really answer the question of how a purely mathematical set of relations can be expressed in the physical realm. It cannot, anyway, be quite right in that general form, since many universes, if actual, prohibit the existence of many other apparently possible universes. Consider the following interesting, and highly relevant, case.

Among possible universes is one universe in which there

exists an all-powerful and all-knowing God, who creates everything else. To avoid confusion, I should point out that I am not now using the world 'universe' to mean a set of physical entities in space–time. I am using it, as a logician would, to mean 'absolutely everything there is', including God, if there is a God. There is, of course, another possible universe which contains no God, but just exists without an external cause. And there is yet another possible universe where there is an utterly evil being who creates a world of intense and endless suffering, just for fun. If all possibilities are actualised, it might seem that all of these universes must exist (it will not make sense to say that they exist either at the same time or one after another, since they have no temporal relation to one another). That in itself, incidentally, completely undermines Atkins' claim that it is much more likely that simple universes should exist than that complex universes should. Indeed, he says that 'only the perfectly simple can come into existence'.[11] But the vast majority of possible universes are very complex, and universes of the greatest possible degree of complexity must tend to be actualised as readily as the simplest. As far as one can see, it is as hard for the simplest universe to come into existence on its own, out of nothing, as for the most complex one to do so.

Atkins' claim that things are, despite appearances, extraordinarily simple is not substantiated by the alleged fact that it is easier for the simple to exist, since there is no such fact. It is just as difficult for one simple thing to have physical existence as for the most complex conceivable universe to have it. If it turns out that this universe is very simple, that is not because such a universe is more likely than a complex one – as we have just noted, this universe is almost infinitely unlikely. It will have something to do with the applicability of very general mathematical propositions to the physical universe, with the elegance, economy, wisdom and beauty of its structure. The mystery of why pure mathematics applies so exactly to the universe is more properly the mystery of why physical elements behave in periodic, regular, mathematically quantifiable and closely correlated ways, such as can be captured by pure mathematical formulae. It is immensely improbable that this should be so. Thus the truth is the very opposite of what Atkins supposes. The simplicity of this universe, its mathematical elegance and integration, far from being quite likely, is immensely improbable. It would be reasonable to accept any postulate that would make it more probable. The postulate that raises its probability to the highest degree is the postulate that some mind – for simplicity, the mind that conceives the infinite realm of possibility – intends to bring into existence a physical realm which actualises a subset of elegant possibilities. That would explain with complete adequacy the extraordinary precision of the Big Bang that began this universe.

It also exposes the fallacy of thinking that every possible world might come into existence sooner or later. Consider the three apparently possible universes just mentioned. Though they at first sight seem to be possible, in fact they cannot all be compossible. Take the first possibility, that there is an all-powerful creator of everything other than itself, God. If that possibility is actualised, then the second and third possibilities cannot ever be actualised. For, on the hypothesis of God, no universe can exist without a cause, and no universe can be created by a supremely evil being, since God would prevent that happening.

It follows that if either the second or third possible universe is actualised, then the first possible universe, and the existence of a creator God, is impossible. It is not the case, therefore, that every possible universe can be actualised. So one cannot think of 'fluctuations of primordial dust' as the actualisation of every possibility, making the existence of this universe inevitable, without the action of a creator. And there is no other sensible way in which the metaphor of 'fluctuation' can be construed, in a regime without time. The necessity hypothesis cannot be supported, and one must revert again to either chance or choice.

CUTTING THE COSMIC BOOTSTRAP

We still have the problem of why any possibility should be actualised at all. Once again Atkins' discussion points us to a solution far removed from the one he actually offers. It points us to the theistic hypothesis rather than to the necessity hypothesis. The theistic hypothesis postulates one necessarily actual reality, the mind that conceives all possibilities. No universe is possible that denies the existence of that mind. We have just seen that the best explanation of the existence of this very precisely specified and elegant universe is that some mind intends it to exist. The ultimate unifying hypothesis is at hand: the all-conceiving (therefore all-knowing) mind is the same mind that actualises one set of possibilities, precisely for the sake of its beauty or goodness. Minds, being conscious, can distinguish between good (pleasant, preferable) states and bad (painful, avoidable) states. A beautiful state causes the best sort of pleasure, and pleasure in the contemplation of beauty is one of the highest forms of goodness. Thus one arrives at the best reason for the existence of this universe: it is chosen for the sake of goodness, the goodness of its contemplation by a mind aware of all the possible states that could exist, with the values and disvalues they carry with them. To put it bluntly, God creates this universe because God wants to enjoy its actuality. It may be that God creates it because God also wants there to exist other, finite, minds which can enjoy its actuality, and which can consciously share in that enjoyment with God. That would explain why this universe is precisely such as to engender conscious beings which come to believe that they can know and love God. This subsidiary hypothesis will be considered further when we come to consider the theory of the evolution of life on earth.

For the moment, the point to emphasise is that all actualisations of a physical realm are highly improbable, and totally inexplicable if they are seen as the genesis of something out of absolutely nothing. Atkins shows that he sees the force of the ancient saying that 'nothing comes from nothing', when he writes that, because the universe comes from nothing, it really is nothing. What else but nothing could come from nothing?

To most of us, however, nothing is more obvious than that the universe really is quite something. And I see great force in that ancient saying, which leads one to suppose that anything actual must either be caused by something actual and with at least as much actuality as it has, or be such that it could not possibly be caused by anything.[12] Once one accepts that view, one has arrived at God. God, being the conceiver and basis of all possibilities, could not be caused by anything (otherwise God would once have been merely possible, and would then have had to be conceived by God, which is absurd). The universe is caused by God, whose knowledge and power is greater than that of any being in all the possible universes that God can cause to be. All things other than God come from God, and God is as far from nothing as it is possible to get, since God is pure and unrestricted actuality.

The theistic hypothesis is that the universe emerges out of the unrestricted actuality of the mind of God, by an intentional act of creation. This hypothesis gives an adequate explanation for the elegance and mathematical beauty of the universe, and raises the probability of its existence to an enormous degree; whereas the hypothesis proposed by Atkins, that 'spacetime generates its own dust in the process of its own self-assembly',[13] is blatantly self-contradictory. It requires that 'time brought the points [non-spatio-temporal entities] into being, and the points brought time into being.

This is the cosmic bootstrap.'[14] It is, however, logically impossible for a cause to bring about some effect, without already being in existence. So if time brought points into being, time must already have existed before the points. And if the points brought time into being, they must already have existed before time. But to say that two things have each existed before the other is a simple contradiction. Since contradictions convey absolutely no information, the cosmic bootstrap turns out to be vacuous. Far from being an ultimate explanation, it says nothing at all. Between the hypothesis of God and the hypothesis of a cosmic bootstrap, there is no competition. We were always right to think that persons, or universes, who seek to pull themselves up by their own bootstraps are forever doomed to failure.

NOTES

[1] Peter Atkins, *Creation Revisited*, p. 3.

[2] Ibid., p. 115.

[3] Atkins says that 'a particle and its anti-particle can be generated out of essentially nothing' (p. 139). If what Atkins says were strictly true, this would be the solution of all our energy problems! In fact, particles and anti-particles resolve into massless energy, but that is far from being nothing. Energy and mass are convertible, but both of them are something!

[4] See E. P. Tryon, 'Is the Universe a Vacuum Fluctuation?'.

[5] See C. J. Isham, 'Quantum Theories of the Creation of the Universe', pp. 49–90.

[6] Atkins, *Creation Revisited*, p. 129.

[7] Ibid., p. 128.

[8] Ibid., p. 133.

[9] Stephen Hawking, *A Brief History of Time*, p. 136.

[10] R. Penrose, *The Emperor's New Mind*, p. 445.

[11] Atkins, *Creation Revisited*, p. 7.

[12] This, of course, is the basic axiom of the classical 'proofs of God'. See Thomas Aquinas, *Summa Theologiae*, 1a, 2, 3, pp. 13f.

[13] Atkins, *Creation Revisited*, p. 143.

[14] Ibid., p. 141.

CHAPTER THREE

Is there any Point? Where the Universe is Going

ENTROPY AND PURPOSE

If the universe is created by God, it clearly has a purpose, and I have briefly construed this purpose as the creation and contemplation of beauty and various forms of goodness, both by God and by finite minds. For Atkins, this is false, because 'everything is driven by motiveless, purposeless decay'.[1] This is another grandiose 'everything' statement, like 'everything can be understood', and 'everything is extraordinarily simple'. Like those other statements, it is clearly false, if taken literally. I am certainly a thing, but if I go to a concert in order to hear Beethoven's Fifth Symphony, I am not driven by motiveless, purposeless decay. I have a definite motive and purpose, and that is to experience some pleasurable and complex conscious state which I take to be worthwhile, just for its own sake. Of course, I know that I am going to die some time, and that my body is in general decaying. I am subject to the second law of thermodynamics, but that in no way stops me having worthwhile purposes.

In a somewhat similar way, the universe as a whole could be involved in an entropic process, passing ultimately to a state of dispersed chaos. There is no need to think that this final physical state of the universe is the purpose for its existence, any more than my final physical state, my corpse, is the purpose of my life. If I have purposes, they will be achieved, not at the very end of my life, but at some point during my life. When I die, I can, if I am lucky, die knowing that I have achieved my purposes. If the universe has a purpose, it will lie not at its physical end, but in the creation and contemplation of worthwhile states during the course of its existence. If such states come to exist, then even if they pass away and cease to be – as they will – the purpose of the universe will have been achieved. Atkins confuses the question about the last state of the universe with the question about its purpose, which is the question whether it was intended to realise some worthwhile states, and whether it did so.

As one looks at the universe, it seems clear that it does realise many worthwhile states – the creation and contemplation of Beethoven's Fifth Symphony among them. The real question is not whether the universe is going to run down, which it probably is, but whether things like symphonies can plausibly be seen as intended at the beginning of this universe. If so, they would be intended by the cosmic mind, God. So one is asking: is this universe such that God could have intended it to bring about many worthwhile states? Could all this have been a result of design and not of chance?

Many of Atkins' own statements suggest a design of enormous wisdom, and so point to a purpose in existence. The balance and precise strength of the fundamental gravitational, electro-magnetic and nuclear forces needs to be exactly what it is if conscious life is to exist. So he says:

> If nuclei were bound together slightly more weakly, or slightly more strongly, the universe would lack a chemistry . . . If the electric force were slightly stronger than it is, evolution would not reach organisms before the sun went out. If it were only slightly less, stars would not have planets, and life would be unknown.[2]

The basic constants of nature need to be exactly what they are to produce life. This, he says, 'may have the flavour of a

miracle'. Well, at least it is just what one would expect if an
immensely wise God wished to produce a life-bearing
universe, if the whole thing was purposive. Whereas it is not
at all what one would expect, if it was a matter of chance.
Every new scientific demonstration of the precision of the
mathematical structure needed to produce conscious life is
evidence of design. Just to go on saying, 'But it could all be
chance' is to refuse to be swayed by evidence. When what is
called 'motiveless decay' leads to every appearance of
purpose, there must come a point at which the rational
observer begins to suspect that it is not motiveless at all. The
process of entropy, after all, is what gives temporal direction
to the universe, and distinguishes past from future.[3] So it too
has a very clear purpose. It is not just a law of the triumph of
decay and purposelessness. It gives a directionality to time,
and out of the eddies and whirlpools of its progress, it enables
local concentrations of energy to form the vastly complex and
intricate structures needed to enable consciousness to emerge.
To see entropy as a process of motiveless decay is to miss its
true character as the generator and enabler of those
concentrations of star-dust, carbon-based life forms, which
come to know and eventually to control, at least in part, the
nature of space and time itself.

THE LAWS OF NATURE: THE RULE-BOOK OF GOD

I have discussed, and discarded, the hypothesis that the universe
originated by chance and the hypothesis that it originated by
necessity. The third hypothesis is that the universe originated in
a free and intelligent choice. On this hypothesis, there exists a
'third thing' which relates eternal and temporal, conceptual and
physical, the necessary and the contingent. We can say that there
does exist a Platonic universe of mathematical entities, including
quantum laws and all sorts of other laws too. But they do not
exist in some half-real realm, neither fully actual nor merely

possible. These entities exist in a cosmic mind. They are the thoughts of a cosmic mind, thoughts that exist by necessity, just as they are. That cosmic mind causes one set of those laws – a supremely elegant, well-ordered set – to apply to a physical reality.

There is a mystery about the fact that the material stuff of the universe obeys general laws. If the whole thing was really random, a matter of pure chance, one would expect that the regularities which the laws of physics describe would change or simply cease to exist after a time. In a universe in which anything at all can happen, the basic laws of physics might at any time just cease to apply. Why should material particles continue to obey such laws as the inverse square law of gravity, or why should the interactions of nuclear particles conform exactly to the restrictions of Schrödinger's equation? The idea of a 'law of nature' was introduced to human consciousness in the High Middle Ages, though it had been formulated, but not used to any effect, by some Greeks. The breakthrough they made was to see that nature was not a playground (or battleground) of many conflicting spirits or demons, but manifested a coherent, intelligible order. In the hands of Isaac Newton, or at least of some of his more incautious followers, the idea of laws of nature was used to construct a special deterministic worldview. According to that view, nature acts in accordance with impersonal laws, which can be mathematically expressed, and which are deterministic in operation. If one could specify the mass, position and velocity of basic particles at any given time, then one could predict exactly what they would be at any future time, by applying the general laws governing their operation. This idea of 'laws of nature' was the fundamental insight that made mechanics, and modern science, possible.

The problem this raises is that of the status of such laws. Do they really exist, and if so, where? How can one be sure that they will continue to apply to nature? Are they just descriptions of what happens, or do they prescribe what must happen? Do they completely govern the future, so that the whole future of the universe is already predetermined at the moment of the Big Bang? I cannot pretend to give a satisfactory answer to these questions. It is enough to point out that they remain highly disputed topics of debate in the philosophy of science. It is noteworthy, however, that Newton himself thought that the laws existed in the mind of God, and that it was a creator God who alone could ensure that physical particles obeyed the laws. While he thought that the laws were deterministic, he did not believe that they alone predetermine the future. The laws hold only as long as no other factors enter in, which might modify them in particular cases. God can modify the laws at any time, for a good reason, and so Newton accepted the occurrence of miracles, for instance.

My own view is that laws, like mathematics, are abstractive; they construct an ideal model of nature, to which nature in its full physical reality does not exactly conform. What one has to do is to isolate some properties (like mass or velocity) which can be measured precisely. These measurable properties are not the only properties things possess, by any means. There are many properties that cannot be measured in discontinuous units, like happiness or thoughts, or even entropy or complexity. To isolate some properties from others is already to construct an abstract model of the world, not to say what the world is *really* like.

Having isolated some measurable properties, one then applies general equations, which describe how the measured properties will interact in a system without interference. Such a model will apply exactly to any physical reality which contains few other properties than those being measured, and which is relatively isolated from other influences. Thus Newton's laws of mechanics apply almost exactly to planetary orbits in space – except for such aberrations as the perihelion of Mercury, which requires relativity. We know they do not apply at speeds near to that of light, though we can find other equations (the equations of special relativity) that do apply in those areas. The laws of mechanics are not so useful in areas where there are many other non-measurable properties in complex interaction – as in social and political relationships. Thus an exact measurement of the mass, position and velocity of billiard balls on a table at the beginning of a game will provide no reliable prediction of where they will be after even two minutes of play. This is not only due to what we now see to be an inherent unpredictability even in Newtonian systems, because of the practical impossibility of describing the initial conditions with utter precision. It is also due to the fact that there are human intentions and abilities involved. Nevertheless, the laws of mechanics will still be useful, for they will enable players to predict what will happen if they hit the ball at a specific angle with a specific force. What they will not do is to predict that any player will be able to pot all the balls in the right order. At that point, psychological and social factors enter the situation, introducing influencing factors which the ideal model of mechanics does not take into account. In this way, one can have deterministic laws and equations in a world not wholly determined by such laws alone. That is the world in which we appear to live.

Even if the laws of physics are not the only factors influencing the future, it is still a remarkable fact that the mathematical constructs of the mind seem to reveal an objective mathematical structure to the physical realm itself. The continuing conformity of physical particles to precise mathematical relationships is something that is much more likely to exist if there is an ordering cosmic mathematician who sets up the correlation in the requisite way. The existence of laws of physics does not render God superfluous. On the contrary, it strongly implies that there is a God who formulates such laws and ensures that the physical realm conforms to them.

THE CREATOR OF THE EMERGENT WEB

No doubt sensing this point, Atkins heroically tries to eliminate laws of nature altogether. He can eliminate the need for a cosmic law-maker and law-giver if he can eliminate laws. His proposal cannot succeed without undermining the whole of modern science. And it turns out that he does not really mean it. What he does is to say that 'the behaviour of things is determined by their nature'.[4] One can avoid having any rules if an entity's nature alone governs its behaviour. So, he says, 'Things happen unless they are expressly forbidden; and nothing will be forbidden.' What, however, is the 'nature' of a thing, which determines its behaviour? The nature of a thing is the set of properties that defines what it is. But how can a set of properties 'unavoidably entail', as Atkins puts it, some behaviour? That can only be the case if these properties already include some dispositional properties, some properties to act in certain ways. Thus one of the properties of light is that it travels as a wave of a particular wavelength and at a particular velocity. That is part of its nature. But that is just another way of saying that there is a rule that light must travel with a precise wavelength and velocity. The rule has not been eliminated; it has just been restated. It is simply not true that 'nothing is forbidden', or that things act with 'perfect freedom'. Light is constrained always to have the wavelength it does have, to consist of photons and to travel at velocity 'c' in a vacuum. Any other action, like travelling faster or slower in a vacuum, is forbidden.

The truth in Atkins' restatement is that the rules are not something arbitrarily imposed on characterless particles, but statements of how things with particular dispositions or propensities interact with one another. That restatement can be helpful in moving one away from a 'clockwork' view of the universe as an impersonal mechanism of parts arbitrarily related by alien rules. The newer view, suggested by quantum and relativity physics, is that the universe consists of fields of forces (like the electro-magnetic or gravitational force), which are in continual dynamic interaction, as different parts of their natures are brought into play at different stages of complexity. In a model I shall deploy later, the universe can better be seen as an emergent web of interacting energies than as a quasi-mechanical clock.

In this new model, there are still laws of interaction which are mathematically statable. Rules have not been eliminated, and the rule-giver has not been pensioned off. It is God who defines the natures of things and ensures the continuity and regularity of their interactions. But now we can see more clearly that God is not merely an external watch-maker. God is the sustainer of a network of dynamic interrelated energies, and might well be seen as the ultimate environing non-material field which draws from material natures a range of the potentialities which lie implicit within them.

The physical reality which even in its first beginnings contains all the potentialities that constitute the nature of things is itself brought into existence contingently, or freely, by creative action. The universe is only one of many possible universes, and the cosmic mind chooses it to exist, and ensures that it conforms to the selected set of laws. This theory enables one to give an explanation for the existence of the universe, in terms of a goal or purpose. The laws themselves exist necessarily, in the cosmic mind. But, as I shall later argue in detail, this universe exists in order to bring about a particular purpose – the existence of conscious, feeling, willing agents which can develop knowledge and action, and share their experiences and actions with one another. The laws of this universe are selected precisely because they are well fitted to realise just that purpose. Matter is not just an eternally existing 'surd' element. It is brought into existence precisely to enable a specific purpose to be realised.

THE MIND OF GOD

This is the theory of creation. God, the cosmic mind, exists by necessity and envisages all possible laws and universes. God is that actual reality in which are to be found all possibilities, and without which nothing at all would be possible. The set of all possibilities clearly includes, as a subset, all possible values that could ever exist in any universe. These are the possible goals of being, the states that God may choose to realise, as being intrinsically desirable and worthwhile. Out of all the possibilities that exist, there is one possible being that actualises supreme perfection, the highest possible set of compossible values. There is, therefore, the best possible reason for actualising this being, if any being at all is to be actualised. But precisely this being is already actual. It is God, a being who is omniscient (who knows all possibles), omnipotent (who can bring about any possible being), and supremely good (who always acts for the sake of goodness). These are precisely the most important qualities of the supremely perfect being. The supremely perfect being, the being that there is the best possible reason for actualising, already exists. Thus, although, of course, God cannot create God, God affirms the divine nature as supremely desirable and as necessarily existing, as the source of all possible beings. It follows that, since God is supremely desirable, God is supremely happy in the contemplation of the divine existence. God is, in the words of St Anselm, that being than which nothing greater, or more desirable, can be conceived or imagined.[5] One possible being, God, exists by necessity, since it cannot fail to exist, as the ground of all possibilities, in every possible world. As for other possible values, God may or may not choose to realise them, as an exercise of that creative freedom which is itself one of the greatest values of the divine being.

This idea of God as the cosmic mind, the necessary basis of all possible states and values, who is, as such, the necessarily actual being of supreme perfection, and the free originator of all other actual states and values, is a supremely simple idea which explains the existence of any universe that is elegantly formed to realise states of distinctive value, states that could not exist in any other universe. God therefore explains the existence of this universe, as a result of the free choice of distinctive values by a necessarily existing and supremely perfect cosmic mind.

So, if one asks what caused the universe, or why it is as it is, the theistic answer is that God brought the universe into being in order to realise a set of great and distinctive values. If one asks what caused God, the answer is that nothing could bring into being a reality which wholly transcends space–time and which is necessarily what it is, in its essential nature, which cannot fail to exist, which is self-existent, the ultimate union of necessary existence and free activity. To fail to grasp such an idea is to fail to grasp what God is. To grasp the idea of God is to grasp an idea of the only reality that could form a completely adequate explanation of the existence of the universe, for God is the only reality which, in being supremely intelligible or comprehensible to itself, explains itself.

It is not constituted by many distinct parts, linked together by chance. Its being is such that all its perfections interpenetrate and are united in one non-dual reality. It is one and indivisible, unlimited in understanding and creative power. Yet diversity is not eliminated; it is transcended by being brought within a greater all-embracing unity. The fully integrated human mind is one, though it contains many thoughts, feelings and purposes. So the cosmic mind is one, though it contains infinite aspects of being within its fully integrated and indestructible unity. The divine unity contains infinite diversity within itself, harmonised and integrated into one all-including totality.

The cosmic mind is beyond speech and utterance, beyond duality and description, beyond being as we can understand it, without beginning or end, without limitation or boundary. There is nothing that can be compared with it, and whatever is spoken of it fails to describe it as it is. It is wrapped in light so bright and blinding that we cannot penetrate it. The ultimately unknown and the ultimately powerful, it is the source of all beings.

That mind gives birth to the universe without being diminished in any way. In its nameless infinity, it always stands beyond, and yet it embraces and enfolds all the finite worlds of time and change. It is infinitely far, and yet no distance separates it from whatever has being. It is the being of all things that have being, which receive it and may think they possess it, until it is taken back to its source.

That mind, from its infinite potency, selects the fundamental laws and limits of this universe, and sets it on its emergent course towards the creation of the galaxies. Yet it upholds every new moment by its constant presence, and without it each moment would fall into nothingness. The One, without name or form, is the infinite depth, the unlimited ocean of being. On its surface all beings come to be and pass away, like foam on a wave,[6] leaving the deep serene and untroubled by their passing. Yet they are parts of it, flung from its infinity. Their power is given but never owned, and they must eventually return to the One who gives it.

NOTES

[1] Peter Atkins, *Creation Revisited*, p. 23.
[2] Ibid., p. 153.
[3] See I. Prigogine and I. Stengers, *Order Out of Chaos*, esp. pp. 177–209.
[4] Atkins, *Creation Revisited*, p. 45.
[5] This can be called the classical definition of God. It can be found in Anselm, *Proslogion* 2, p. 117.
[6] See J. A. Wheeler, *Gravity and Spacetime*.

CHAPTER FOUR

Darwin and Natural Selection

EVOLUTION: THE COSMIC PROCESS

From an initial singularity, the universe exploded into movement and time, expanding to create three-dimensional space and rapidly cooling to form the elements of hydrogen and helium, which make up the great majority of elements in this universe. The rate of expansion was precisely that needed to produce the exact ratio of light and heavy elements which would alone make possible the origin of galaxies and star systems.

Stars formed from the hydrogen and helium atoms, and as they aged, they exploded to form carbon, the basis of all life-forms. From the destruction and the dissipation of energy, more complex forms came into being. From the formation of planetary systems, various elements such as carbon, hydrogen, nitrogen and many others combined to form the very finely balanced environment capable of originating and sustaining organic life-forms. From the standpoint of modern science, the whole history of the universe is a history of evolution, of the development of complex structures out of simple elements, in accordance with in-built principles of interaction, or 'laws of physics'. On this planet, that process of evolution has continued, as simple atoms of carbon, hydrogen, oxygen and other elements built up and developed into the millions of life-forms that now exist on earth. That such a process has taken place is one of the foundations of modern science, and the idea of evolution explains many otherwise puzzling features of our world in a very elegant way.

Charles Darwin did not invent the idea of evolution, by

any means, but he has been the most influential exponent of
the idea, particularly with regard to the evolution of organic
life. He proposed a particular, and tremendously influential,
account of how evolution works, of how all living organic
forms have developed from some simple primeval ancestor.
His theory – the theory of natural selection – has possibly
changed our view of human existence more than any other
theory in the history of human thought. This theory states
that living beings reproduce themselves with very slight
variations over very many generations, and are engaged in a
continual struggle for life, in the face of competition with
other living beings and with the forces of nature. 'Owing to
this struggle for life, any variation, however slight . . . if it be
in any degree profitable to an individual . . . will tend to the
preservation of that individual, and will generally be inherited
by its offspring.'[1] Darwin did not think this was the only
principle accounting for evolutionary change, but he did
think that it was by far the most important one.

It is a very simple hypothesis. It requires only that slight
mutations occur in the replication of the genetic codes, and
that there is a great number of replicating organisms
competing for survival. As Darwin put it most forcefully,
'There is one general law, leading to the advancement of all
organic beings, namely, multiply, vary, let the strongest live
and the weakest die.'[2] This gloomy view of a continual battle
for survival between egoistic organisms was suggested to
Darwin by his reading of Malthus, who had predicted that
the inevitable tendency to over-population could only be
remedied by repeated wars and plagues if life on earth was to
survive.

The theory of natural selection is a simple and
extremely fruitful one. To many biologists, it provides the key
to explaining the complexities of animal life. It is a scientific
theory, and so it may seem not to conflict with religious
beliefs at all. Certainly, those religious believers who rejected

Darwin's theory because they thought it conflicted with the account of creation in the Book of Genesis have been pretty thoroughly discredited. They were wrong both about how to interpret Genesis and in thinking that evolutionary theory was unimportant to modern science. Augustine saw long ago that Genesis was not to be taken as a literal description of physical facts.[3] There is no greater difficulty in thinking that God brought living forms into being through a gradual process of evolution than in thinking that creation happened all at once. In many ways, the evolutionary account is more impressive, since the development of complex and integrated forms which can support consciousness and agency out of simple atomic elements suggests an immense and patient wisdom and a purposive guidance underlying the whole process. That the whole cosmos has developed from simplicity and unconsciousness to complexity and self-awareness is a foundational view of modern science. It is not only biology, but cosmology, physics and astronomy that presuppose a general evolutionary account of the cosmos. Such an evolution from a state where no values are apprehended to states in which values can be both created and enjoyed gives an overwhelming impression of purpose or design. There is thus every reason to think that a scientific evolutionary account and a religious belief in a guiding creative force are not just compatible, but mutually reinforcing.

Nevertheless, Darwin's early opponents were right about one thing. Some interpretations, at least, of the hypothesis of natural selection do conflict with the theistic hypothesis on three main counts. First, they see the process of evolution as a non-purposive, fortuitous (if that is the right word) accident. It is not, of course, strictly necessary to take this interpretation. One could hypothesise that God uses natural selection in order to bring about a special set of purposes, which could be obtained in no other way – for instance, purposes that require a combination of order and openness in the development of the universe.[4] So this is an argument between one interpretation of natural selection and one interpretation of religion, not between natural selection and religion as such.

A second area of possible conflict is that some interpretations of natural selection see evolution as a ruthless struggle for survival, in which the strong inherit the earth, and the weak are exterminated. If this is a complete account, it clearly conflicts with the theistic view that love and humility are among the most important goals of human existence. We shall see that, here again, natural selection is not as such committed to the ruthless interpretation, and accounts quite well for co-operation and altruism, as well.

Third, some interpretations of natural selection see the emergence of mind, value, purpose and creativity as, at best, a helpful adjunct in the battle of genes for their own survival, and as, at worst, a sentimental obstruction to the battle for survival which is most efficient when it is unimpeded by conscience or by attempts to act on supposedly rational principles. There is a conflict here with any religious view that thinks that the existence of finite conscious beings is the goal of the evolutionary process, whether it is brought about by natural selection or not.

The arguments for these three interpretations of natural selection are quite remarkably weak, and do not carry the authority of the scientific theory of natural selection, which remains a powerful explanation for a wide range of biological phenomena. I hope to show that the arguments do not survive close examination, and are weaker than a theistic interpretation of the same range of facts.

NATURAL SELECTION: AN UNCERTAIN PROGRESS

Is the emergence of sentient, rational, moral agents from

simple unconscious virus-like cells purposive? In asking this question, one is asking if it happens by an efficiently structured process which issues in a state with intrinsic value or desirability. Any such process has a prima facie claim to be purposively ordered. On this criterion, evolution looks clearly purposive. Beginning with a state in which no values or worthwhile states are apprehended at all (where there is no consciousness), and proceeding by the operation of highly elegant physical laws, evolution has arrived at a state in which highly structured, self-replicating organisms know, feel and act. They are capable of producing and enjoying states of intrinsic value, desirable for their own sakes alone. If there are such things as intrinsic values at all (and I shall later argue that there are), this is just the sort of process a being could intelligibly create, in order to realise the purpose of bringing about the emergence of such states of value, in accordance with general intelligible laws, from insensate matter. Given the hypothesis of God, together with the postulate that God intelligibly desires the existence of intrinsic values of that sort, and will act to realise some set of those desires, the existence of a universe like this follows with virtual certainty.

The resolute Darwinian typically claims that the hypothesis of God is unnecessary, and that the simpler hypothesis of natural selection alone accounts for all the facts quite satisfactorily. This, however, is more a blunt, and as yet unsubstantiated, assertion than an argument. Any assertion that the hypothesis of natural selection can account for *all* the facts is a remarkably bold claim, when so many facts still remain unaccounted for, even in the realm of evolutionary biology. It is doubtful whether the hypothesis even accounts for all the *known* facts in an entirely adequate way. What the theory of natural selection proposes is that mutations which are not biased towards any 'development' or 'improvement' will occur in such a way that some mutated organisms will be more efficient at reproducing and will thus tend to survive in the constant struggle for life which eliminates most of the others. After many generations of such mutations, the life-forms that result will, by definition, be the best survivors and reproducers.

In what sense does this principle *explain* the emergence of complex conscious life-forms from much simpler unconscious organisms? It certainly cannot guarantee that such an emergence will occur. Perhaps the mutations that occur will be too large, too small or too arbitrary to generate viable complex organisms.[5] Perhaps mutations in the chromosomes that carry the mechanisms of heredity will not correlate with the other structural changes in an organism necessary to produce a more efficient predator or survivor. Perhaps the environment will eliminate a whole class of organisms, no matter how superior in complexity they are (as the dinosaurs were wiped out, probably by some environmental change or catastrophe, whether a comet or something else). Perhaps the emergence of consciousness will have an adverse survival value (as would be the case if, for example, the reproductive act happened to cause consciousness of intense pain). The number of things that can go wrong with the process is enormous. So the principle of natural selection does not seem to make the development of conscious life inevitable (as a good theory in physics, like the theory of gravitation, makes its effects inevitable, other things being equal). It does not even make such development more probable than not. For there is nothing in the principle to guarantee that the 'right' type of progressive mutations will ever occur, that the environment will favour them, or that in a struggle for life, the more complex organisms will be favoured.

Biologists disagree about whether natural selection makes increased complexity probable or improbable. W. McCoy, for example, has argued that 'evolution is a process of divergence and wandering rather than an inexorable progression towards increasing complexity'.[6] This, one might think, would make the emergence of great complexity improbable, since structures normally tend to degrade and disintegrate through random interaction. Arthur Peacocke, however, speaks of a 'propensity for increased complexity',[7] some inherent weighting of evolutionary change which favours complexity. This, of course, would make the development of complexity probable, perhaps inevitable.

The problem is to see what the cause of such a bias to complexity could be. There might be some weighting of the mutational process, to ensure that more complex mutations occurred. Then competition for scarce resources might favourably select more complex organisms. Obviously, competition will select entities more efficient at obtaining energy (food), at defeating competitors and at replicating themselves. Insofar as complexity increases such efficiency, it will be favoured. The qualities selected for are efficient energy-consumption, fighting ability and reproductive capacity. If an organism emerges which is just a little bit better at food-gathering, fighting and replicating, it will be naturally selected.

What we cannot tell in advance is whether such an organism will emerge by mutation, or what exactly it will be like, or whether it will continue to be favoured by the environment. For example, if one food-source becomes extinct, organisms that can use another food-source will be favoured. The aggressiveness that leads me to eliminate enemies may destroy my species too. Or a natural disaster may wipe out my offspring. In all these cases, what seemed in advance to make for 'fitness' for survival may turn out not to do so.

Because of such factors, increase in complexity cannot be guaranteed. And if one looks at what Dr Peacocke means by a 'propensity', it is only that 'there has over biological evolution as a whole been an overall trend towards and an increase in complexity'.[8] Since we only have one planet to go on, this simply means that the process has happened once. It does not mean that it is probable (that it would happen on most worlds) at all. The upshot seems to be that the principle of natural selection cannot make any particular evolutionary path much more probable than any other.[9] It can, however, suggest that, *if* all the relevant causal and environmental conditions are right (the 'right' mutations occurring in the 'right' environment), organisms with a certain degree of complexity and organisation are likely to be selected. That, in turn, is a necessary condition of the existence of consciousness, which is likely to be selected if it ever occurs.

This means that the evolutionary process depends entirely upon the right sort of mutations occurring in the right sort of total environmental conditions. I think that what Peacocke, a theist, means by 'propensities towards increase in complexity, information-processing and storage, consciousness, sensitivity to pain, and even self-consciousness'[10] is that the total causal and environmental conditions of evolution are 'built-in and intended by God their creator'[11] to produce exactly such qualities. Here it must be kept in mind that the 'total causal and environmental conditions' must include the being of God itself, the ultimate cause and goal of the universe. Theists are not thinking of a universe that is 'natural' in the sense that it is unaffected by God in any way. They are not thinking, either, of a universe that is constantly being interfered with by a God who keeps breaking the laws of nature. Theists are thinking of a universe that is continually kept in existence by a God whose nature inevitably and continuously affects and guides all the processes within it, so that in their own proper nature they express the divine purpose of creation. In this sense it is only the existence of God that can explain the propensity to complexity and consciousness that seems so clearly present in evolution.

HOW MUCH CAN NATURAL SELECTION EXPLAIN?

On grounds of natural selection alone, prescinding from any idea of a creator with a purpose, there does not seem to be any reason to expect such a propensity towards the development of complexity. Natural selection does not properly claim that more complex organisms will be favoured. It simply says that, in a struggle for scarce resources, there will have to be some winners and some losers. The winners will be called 'the fittest', just because they have survived and the others have not. But one cannot tell in advance who the fittest will be. A good example is the human species. One might think humans are the best fitted to survive on earth, because they are conscious, intelligent and capable of changing their environment. But it is perfectly easy to envisage that, precisely because of these abilities, coupled with a basically warlike nature developed precisely through the struggle for survival, humans will wipe themselves out quite soon. Then something else – probably ants, well suited to survive the radiation produced by nuclear fall-out – will survive, and they will prove to have been the fittest. The point is that we do not know who the fittest will be, in advance. It is just a matter of who actually survives, not who ought to survive. It is, after all, a matter of luck whether conscious complex organisms will manage to survive better than insentient simple ones.

In fact, all the principle should really claim is the rather negative point that some mutations do not get eliminated in the struggle for life. It is not entitled to claim that the mutations that survive do so *because* they have a high survival value (as though that could be established independently of seeing whether or not they happen to survive). It is easy to think of examples of survivals which have not been eliminated, but which have no survival value at all. The human appendix is one organ that continues to be replicated, even though it has lost any survival value it once had. If hard-line materialists – who think that consciousness is just a by-product of brain activity – are right, the existence of consciousness itself is another. For materialists, consciousness plays no role in the causal process of the natural world, so it is strictly irrelevant to survival. Yet it has survived rather well in our world. In other words, one cannot explain the existence of all properties of organisms by referring to their survival value. It looks as though here, too, the hypothesis does not account for *all* the biological facts.

I am not making the absurd claim that the theory of natural selection is useless. If one wants to know, as Darwin did, why there are similar but variant species in different places, the hypothesis that the accumulation of many small mutations led to such variations is convincing. Very often one can see how the process operates. In arctic climates, bears that happen to have white coats will be better protected than brown ones – if the climate does not change, if there is sufficient food to eat, if there are no predators that particularly like white fur, if the mutation to white carries no other disadvantageous gene with it, and if there is no even better mutation amongst brown bears – perhaps to having poisonous fangs, for instance. The process is extremely 'iffy'.

Nevertheless, the theory works well if, using hindsight, one looks back at what has actually happened. One can say that human life-forms are very good at surviving in different environments and at reproducing, so that they have achieved dominance on this planet simply by their superior efficiency in these respects. One can explain many apparently strange human characteristics by pointing to their value for survival at various stages of evolutionary development. Within its limits, natural selection is an illuminating idea. Even so, it seems a rather drastic procedure to pretend that all distinctive human characteristics (especially the development of consciousness, morality, rationality, science and art) can be adequately explained by showing that they were conducive to more efficient domination or reproduction. Such attempted explanations (in human sociobiology, for example) are highly controversial, and have never been provided in any convincing detail. They are at best proposals for explanations that might be provided in future, rather than actual explanations of specific details. The programme of explaining characteristics of dominant life-forms in terms simply of survival value is controversial and highly speculative.

One is reminded of Karl Marx's attempt to explain all human culture in terms of economic forces of production and exchange. The attempt had a certain simplistic appeal, and could lead one to look at many cultural phenomena in a new, usually more sceptical, light. It had the advantage of iconoclasm, a pleasing shock-effect which derives from an adolescent sense of mocking traditional values. But the trouble with the shock of the new is that it quickly becomes the familiarity of the old. In a more reflective mood, the Marxist explanation turned out to be almost vacuous, in detail. Simplicity of explanation should not be bought at the price of eliminating the most distinctive features of the very phenomena one is trying to explain.

So the proposal to explain something like the production of Mozart's symphonies by appealing either to economic necessity (the Marxist explanation) or fitness for survival (the Darwinian explanation) looks, in the clear light of day, wildly inadequate. Certainly, musical ability is a property that may be genetically inherited and which has emerged and survived in the human gene pool. But to try and explain its existence by showing that it had survival value requires a very high standard of story-telling or myth-making ability. Even when one has told such a story, it is pretty clear that there are many other equally good stories one could tell.

Darwin says at one point, 'I can see no difficulty in a race of bears being rendered, by natural selection, more and more aquatic in their structure and habits, with larger and larger mouths, till a creature was produced as monstrous as a whale.'[12] There is no difficulty, either, in imagining a race of whales gradually becoming more and more land-loving, perhaps in their pursuit of seals, with fins developing into legs and bodies shrinking to a more mobile size, till a creature is produced as furry as a bear. That is exactly the trouble with this modern-day scientific myth. With the aid of hindsight, it can explain absolutely anything. This leads one to suspect that it is such a general form of explanation that it needs the addition of many other forms of explanation if anything like an adequate explanation of the evolution of conscious rational beings is to be attempted.

Even as a hindsight theory, natural selection seems on the one hand to be much too over-simple and monothematic to be able to give an adequate account of the development of all the diverse qualities of complex sentient life-forms. On the other hand, it is too vague and flexible to provide a satisfyingly specific explanation of the evolutionary process, though it must be regarded as a powerful part of any such explanation (which is all that Darwin himself wanted).[13] There is a place and a need for other forms of explanation, and amongst these might well be the sort of purposive explanation that theism can provide.

PROGRESS IN EVOLUTION: DARWIN'S DOUBLE-MINDEDNESS

The theory of natural selection is even more inadequate as a predictor of what life-forms are liable to be produced by the application of the theory to primitive self-replicating organisms. That does not stop it being a good scientific theory. After all, palaeontology and geology contain plenty of good scientific hypotheses which do not yield accurate predictions. But it does throw doubt on any claim that natural selection is a completely adequate explanation of all the mysteries of evolution.[14] A *completely* adequate scientific theory should make the occurrence of the specific events that actually happen highly probable. Newton's laws, as a paradigm case, make the occurrence of events in a mechanical system free from external interference not only highly probable but virtually certain. The strange fact about the theory of natural selection is that, according to Darwin in at least one of his moods, it does not make the emergence of rational life-forms certain, or highly probable. Indeed, it makes such an occurrence very highly improbable. That is, after all, a rather odd thing for a scientific hypothesis to do.

According to the theory of natural selection, mutations are random; that is, they have no in-built tendency to develop in any particular direction. Darwin says, 'I believe in no law of necessary development.'[15] He rejects with some vehemence any idea that there may be a purpose in the evolutionary process: 'Nothing can be more hopeless than to attempt to explain . . . by utility or by the doctrine of final causes.'[16] There is no reason why things should 'progress' at all. Yet Darwin is double-minded about this, and often writes in quite a different vein: 'As natural selection works solely by and for the good of each being, all corporeal and mental endowments will tend to progress towards perfection.'[17] He even personalises natural selection, as a sort of quasi-providential power: 'What limit can be put to this power, acting during long ages and rigidly scrutinising the whole constitution, structure and habits of each creature – favouring the good and rejecting the bad? I can see no limit to this power, in slowly and beautifully adapting each form to the most complex relations of life.'[18] In this mood, Darwin speaks of progress without any obvious embarrassment: 'There is no logical impossibility in the acquirement of any conceivable degree of perfection through natural selection.'[19] But that is just the trouble. It is logically possible that any degree of perfection may come about. But it is just as logically possible that no perfection will come about at all. The theory will be able to explain absolutely anything that happens, simply by saying that it was obviously the result of the struggle for survival. That must make one suspicious of any claim that the theory gives a complete explanation of every specific thing that happens.

All the theory can say is that some mutations will be grossly disadvantageous, so will be eliminated. Some will give a slight advantage – though one cannot tell in advance what form that advantage will take, except that it will increase reproductive efficiency and ability to survive in whatever environment happens to exist. Even this, however, gives little clue about what life-forms may pass the survival test. As Darwin writes, 'Probably in no one case could we precisely say why one species has been victorious over another in the great battle of life.'[20] This really amounts to saying that there may be some explanation for the course evolution has taken on earth, but we do not know what it is. It may seem that organisms that develop a capacity for sensitivity to their environment (knowledge) and ability to change it (movement and purpose) will be liable to survive best. But, bearing in mind that there is no question of organisms 'trying to survive', and thus devising optimal strategies for survival, there is no particular reason in the principle of natural selection itself why such capacities should develop by mutation in the first place.

Since natural selection can only operate on the mutations offered to it in specific sets of environmental conditions, it may well be that the processes of mutation and the nature of changes in the ecosystem are more important in explaining the evolution of life than the principle of natural selection, important though that undoubtedly is. Eldredge and Gould have developed the hypothesis of 'punctuated equilibria', according to which long periods of gradual mutation are punctuated by episodic events in which large, fast, saltatory genetic changes (i.e. changes by large sudden jumps) occur in conditions of relative genetic isolation.[21] Such changes occur before any selectional control, though of course they are subject to natural selection once they exist. This means that they play a major explanatory role in the evolutionary process. This, in turn, suggests that while natural selection itself plays a necessary part in understanding, it is by no means a complete explanation of evolution.

It seems quite plausible, especially on a gradualist view of phyletic change, that some form of virulent virus may be most efficient at exterminating other organisms and at replicating itself in many mutating forms, all of them unconscious and without any formulated purpose. Small genetic advantages are liable to be swamped in large genetic populations, and to be wiped out by different sorts of advantage incompatible with them. The virulent virus will survive in a condition of small, fluctuating mutations, efficiently exterminating anything more complex, which would, if established, assume a position of dominance.

On a punctuated equilibrium view, some individuals might suddenly accumulate huge advantages (as apparently with the species *Homo sapiens,* which took over the whole habitable world in about 15,000 years). This could only happen if, as with the development of a hominid brain, the saltation carried a large survival advantage. Stephen Gould himself does not see evolution as a purposive process, and so this seems, if anything, even more improbable than it is on the theory of phyletic gradualism. It is easy to see, however, that what seems extremely improbable on a 'chance mutation' theory would become highly probable on some view (like a theistic view) that builds a tendency to develop conscious beings into the structure of the material universe itself.

It is one thing to say that beings that are fully conscious of their surroundings and can bring those surroundings under their control will survive well. It is quite another thing to say that such beings will probably come into existence as a result of the repeated application of a completely blind and non-purposive process of organic mutation and replication. The former suggestion is plainly true. The latter suggestion, however, seems to be wholly improbable. Perhaps it is no more improbable than anything else that may happen by chance in this universe. But there is nothing in the theory of natural selection to make it probable that sentient life-forms will develop by mutation (that mutations in the direction of sentience will occur at all), or will survive best in potentially hostile environments (that there will be environments that tolerate them). According to natural selection, in other words, the emergence of sentient life may be no more improbable than many other possible consequences of the repeated application of basic laws of nature. But it is still highly improbable, a fluke of nature, not a predictable outcome.

THE DIFFERENCE GOD MAKES

The theory of natural selection cannot predict that sentient life-forms will come into existence, and in fact it makes their existence highly improbable, even though possible. This would not normally be taken as a sign of a completely explanatory scientific theory. It would be entirely reasonable to conclude that the hypothesis of God – of a cosmic mind which sets up the processes of mutation so that they will lead to the existence of communities of conscious agents – is a much better hypothesis than that of natural selection. For on the God hypothesis, the development of sentient life-forms from simple organic molecules is very highly probable; whereas on the natural selection hypothesis, such development is very highly improbable.

It is open to a theist to argue that natural selection plus the basic laws of physics does make the development of sentient life-forms probable. In that case, one could hold that God has designed the basic laws so that, in the long run, in one way or another, conscious beings would come to exist. One would see natural selection as the way in which God works, without interference in the laws of nature, to realise the divine purposes in creation. God would not be needed to explain why natural selection moves in the direction it does, when it could easily have moved in some other direction (or in no direction at all). But God would still have an explanatory role, in providing a reason why this set of physical laws exists, and in assigning a goal (of conscious relationship to God) to the process of evolution.[22]

For such a view, God and natural selection would not be competing hypotheses. God would be the ultimate cause of the finite causal processes embodied in natural selection, but would not interfere in those finite causal processes, as an additional cause. I am not happy with accepting this otherwise attractive view. I am not convinced that the principle of natural selection alone makes the emergence of rational beings probable. Of course it is *possible*. Almost anything is possible, where there are no fundamental metaphysical constraints on the sorts of things that might exist. But I do not think that what is possible is bound to happen, sooner or later. Taking natural selection alone, it seems to me highly unlikely that rational beings should ever come to exist in a universe like this, just as it is highly unlikely that this universe should exist at all.

To make it likely that rational beings should emerge, there would have to be some 'weighting' of the probabilities

of events occurring which would make the emergence of
rationality inevitable, sooner or later. It is very hard to see
what physical weighting there could be, and in my view the
suggestion of such a weighting is not preferable to the
hypothesis of divine causal influence.[3] A physical weighting
ought to be physically detectable, in the physical structure of
molecules perhaps, and it has certainly not been detected. So
my view remains that, while natural selection can explain
why more accurate information-processing organisms should
be preferentially selected once they come to exist, it cannot
make the initial mutations that generate more accurate
information-processing systems more likely to occur than not.
Also, it cannot make it likely that the environment that
permits the operation of natural selection will remain tolerant
of any life-forms at all, much less tolerant of more complex
forms. For these reasons, I regard evolution by natural
selection as a much more insecure and precarious process
than seems compatible with the theistic idea of a goal-
directed process. With this judgement, of course, I would
expect neo-Darwinians like Richard Dawkins to agree whole-
heartedly. I cannot see any way, in terms of physical causes
and processes, of remedying this precariousness. In other
words, the process, while possible, is too unlikely and
precarious to be set up and then left strictly to itself by a God
who intends rational beings to come into existence. In this
sense, a continuing causal activity of God seems the best
explanation of the progress towards greater consciousness
and intentionality that one sees in the actual course of the
evolution of life on earth.

How can one conceive such a continuing causal activity of
God? Most theists would wish to speak of God as having some
causal effects on the world. They accept the possibility of
conscious human relationship to God, and of God being
responsive to human acts and prayers. They wish to speak of
God revealing certain facts, or guiding certain human lives,

and in general causally interacting with the human world. The form of such causal interaction must be different from that of normal physical causality, since God is not a finite or physical cause. Yet it seems clear that most theists would want to say that the existence of God makes the course of human life different than it would have been had there been no God. That entails that the processes of the world must be causally influenced by God, to proceed in ways that they would not, or might not, have proceeded in without God.

One does not have to think of this as a matter of there being a set of deterministic laws of physics or biology, which God interferes with occasionally to push things in the right direction. I prefer to use the idea, canvassed by Arthur Peacocke, of 'top-down' or 'whole-part causation',[24] whereby, roughly speaking, the nature of a complex whole influences the nature of its parts. For a theist, the ultimate complex whole consists of the universe and God. God is the ultimate reality, constantly holding the universe in being and determining its general nature. Some theists (more properly, 'deists') believe that God sets the universe up at the beginning, and then has nothing more to do with it. But a more satisfactory notion of creation is that God at every moment sustains the universe, so that every moment is a moment of creation.

That entails that God either actively or passively determines what the universe shall be at every moment. God determines a state actively if it exists by the direct intention of God. God determines a state passively if it exists as a result of finite causes which God allows to operate, without directly intending it. God will not completely change the nature of the universe from one moment to another, or there would be no intelligible principles of change and no coherence in the universe at all. But God is certainly not constrained to let creation at one moment be *wholly* determined *solely* by what preceded that moment, plus a set of utterly universal and general laws. God is free to respond in new ways and to initiate new courses of action. The very fact that God has particular intentions for the future of the universe entails that the nature of the parts of the universe will be affected by those intentions. There will be a 'top-down causality', running from God, the ultimate sustaining and creative power, to all parts of the universe, which have their own finite powers and capacities.

This causality might be best envisaged as what the mathematical physicist and philosopher A. N. Whitehead called a 'lure' towards greater development or fulfilment of positive potentialities.[25] Divine causality will be universal in time and space. It will be a constant influencing factor which does not interrupt or 'interfere with' the proper powers of finite objects, but which co-operates with those powers to guide them towards fulfilment of the divine purpose. The theist is able to postulate an ultimate spiritual environment for the physical universe, which will exercise influencing constraints on the way it unfolds its inherent potentialities. This is one way of understanding the most general causal influence of God on the course of evolution.

THE ACTIONS OF A SIMPLE GOD

One may think of God as having a universe-long intention to bring conscious beings into a community of freely chosen loving relationships. This intention will shape the initial laws of the universe and the emergence of more complex possibilities within it. In general, God will exert the maximum influence for good compatible with the preservation of the relative autonomy of nature and its probabilistic laws, and with the freedom of finite agents. God's causality will be physically undetectable, since the divine influence is not a quantifiable property, like mass or energy. But it will be an ultimate parameter or constraint, which affects every part of the physical universe in a different way, depending on the context and organised complexity of each part.

The existence of such divine causality entails that a completely determinist account of physical causation cannot, in principle, be given for all physical processes. Such an account cannot be given anyway, even in strictly Newtonian terms. As physicist John Houghton says, 'Initial conditions can never be specified absolutely precisely, so that the predictability horizon represents a fundamental limit to our ability to predict.'[26] When one tries to specify initial conditions with absolute precision, one runs up against the Heisenberg uncertainty principle, which precludes our simultaneous knowledge of all relevant physical variables. Whether this is due to a fundamental uncertainty in nature, as most physicists think, or not, it certainly places limits on our ability to predict. And such uncertainty must extend to the large-scale differences that result in so-called 'chaotic' regimes, where fluctuations at a micro-level lead to marked transitions in the macroscopic state of a system.

Since completely deterministic prediction is now said by most physicists to be impossible in principle, it must remain an open question, as far as physics is concerned, whether or not physical laws operate deterministically, without 'interference' from a divine source. I dislike the word 'interference', when speaking about the causal influence of a God who constitutes physical systems to be what they are. But, as far as physics is concerned, it is quite possible for God to influence the outcome of physical events, within certain limits, in ways undetectable by us. This is true, whether or not quantum uncertainty is a fundamental feature of physical reality. The only objection to a continuing divine causal influence on the physical universe is thus a hangover from an outmoded Newtonian determinism, which was always more properly a philosophical theory than a scientific one.

The determinist theory held that it would be irrational, or less than optimally rational, for a perfectly wise God to intervene in laws God had personally devised. For that would show some imperfection in the original design. Suppose, however, that one adopts a more personalist model of God, as a being who intends that there should be an interactive relationship between finite created persons and God. That would entail the existence of causal relations between such persons, in all their physical complexity, and the divine being. There would have to be a real causal influence of created persons upon God, and of God upon the physical universe. Given such a personalist model, there would have to be a causal influence of God upon those parts of the physical universe that constitute the brains and central nervous systems of persons, and upon those prior processes that give rise, through evolution, to the existence of such central nervous systems. In other words, a rationally coherent idea of a personal creator God entails that the laws of physical nature are not fully deterministic. Rather, all events in nature will be influenced in some way by the goals and intentions of God. What modern physics helps to show is that such influence cannot be ruled out by the discoveries or presuppositions of science, and that the influence of God will nonetheless, in normal circumstances, be physically undetectable.

The hypothesis of a continuing and universal divine causal influence may be seen as a generalisation of an understanding of God's present relation to humans extending to the universe as a whole. If God can be seen as having a responsive and creative relationship to parts of the physical universe (human beings) now, then there can be no a priori objection to God's having a relationship of that general type to the physical universe as a whole. In fact, rational consistency may imply that God would not wait until the advent of humans to have a providential relation to the universe.

One can now account more plausibly for the 'weighting' of probabilities towards more complex and sentient life-forms. There is no physical mechanism that produces such a bias. Yet it is not left entirely to chance. It is the being of God, which alone sustains the universe, that exercises a constant causal 'top-down' influence on the processes of mutation and natural selection, and guides them towards generating sentience and the creation and apprehension of intrinsic value. Such continuing action of God should not be thought of as an 'interference' with otherwise totally explanatory laws of nature. For it is the most general parameter which governs the processes of temporal change that constitute laws of nature.

Many theists will wish to speak, in addition, of 'miracles' as points at which physical structures transcend their normal modes of operation, having been united in a special way with their spiritual basis and goal. Insofar as this is the case, miracles too will not simply be freak events, arbitrarily caused by God. They are not well described as 'violations of laws of nature', a description chosen by the philosopher David Hume precisely because it helped to make miracles seem immoral and irrational. Rather, miracles are occasions when normal physical regularities are modified by a more overt influence of the underlying spiritual basis of all beings. From a theistic viewpoint, such modification will show finite things in their true relation to their infinite ground. It will not be an arbitrary breaking of rational and self-contained laws. Thus miracles have their own internal rationality, which can probably only be perceived by us when the totality of the cosmic process is completed. Miracles, however, have a religious purpose, not a scientific one. They tend to be grouped around prophetic figures, whose conscious relation to God makes possible dramatic disclosures of the divine nature and purpose, in and through particular events. One may be fairly confident that the processes of biological evolution do not require miracles in order to produce rational agents, and I am not suggesting that they do.

When, from this point on, I speak of 'natural selection', I shall be assuming an interpretation of natural selection that regards it as having no bias, whether inherent or divinely prompted, towards the emergence of rational agency. I shall view it as a principle that operates without any bias towards the emergence of certain sorts of organism. Taken in this sense, the basic difference between the hypothesis of natural selection and the hypothesis of theistic design is that, for natural selection, the only reason why organisms of a certain kind develop is that they happen to have survived, though they might very easily not have developed at all, or have survived. For theism, on the other hand, life-forms develop in order that conscious beings, capable of producing and enjoying distinctive sorts of value, should come into existence.

Insofar as Darwin eliminated purpose from the process of evolution, one might claim that he made the whole process much simpler (it has one fewer entity to consider, having eliminated God). Unfortunately, he also made the process very much less probable, and that is not a good thing for a purported explanation to do. Even the appeal to simplicity is not as simple as it sounds. Darwin is left, like the reductive cosmologists, with a number of unexplained laws of nature, with no explanation of how they happen to integrate so as to produce stable life-forms, no explanation of the vast improbability of the course of evolution, and no explanation of how mind, awareness, purpose and value can be parts of this process at all. The simplicity his theory provides might be called an 'exclusive simplicity', one which ignores and excludes virtually all the complexity and diversity, the beauty and value of things. It is easy to get simplicity in this way, by simply eliminating all complexity. We have seen that Atkins tries to do this in cosmology. But one does not really explain anything by eliminating it, and then claiming that there is nothing to explain.

There is another sort of simplicity, 'inclusive simplicity', which embraces and includes complexity and value in one integral unity of primordial fullness. In this sense, God, the one source of all being and value, of freedom and necessity, of unity and diversity, of matter and consciousness, is the ultimately simple reality. Exclusive simplicity is radically impoverished, since it is forced to discount all the richness of experience and conscious life. However, the rich diversity-in-unity of the inclusive simplicity of God is able both to range over the whole array of diverse features of the universe and to unite those features by appeal to the simple idea of one necessarily existing perfect creator of all finite things. The simplicity in question here is the simplicity of an elegant principle which unites the elements of the complex in a coherent and integrated way. It cannot afford to ignore the elements of the complex. So the simplest explanation is the one that can include the whole range of complex elements within one integral and harmonious scheme. Some of the most important elements of organic life are consciousness, value, a sense of moral obligation, beauty and culture. Any simple principle which is able to integrate these elements with the basic facts of the physical universe must embrace elements of purpose and freedom as well as of chance and necessity. It is because of this that God is actually a simpler hypothesis than natural selection alone. The idea of 'one necessarily existing creator of all things for the sake of goodness' is a much simpler idea than the postulate of natural selection alone, which leaves unexplained and therefore unintegrated all the phenomena of consciousness, freedom and purpose that characterise higher organic life-forms.

THE WAR OF NATURE: DARWIN'S GLOOMY VIEW

So far I have focused on the surprising emptiness of
explanatory force in Darwin's theory of natural selection –
surprising, in view of the great claims that are made for it by
some speculative biologists. But another important feature of
his theory is its reliance on a particular set of metaphors for
understanding the natural world. These are metaphors of battle,
struggle and the survival of the fittest. Is it really an objective,
dispassionate and non-evaluative view of nature which sees it as
a constant battleground of competing powers, each ruthlessly
seeking survival at any cost to others, 'red in tooth and claw'?
Darwin writes, 'From the war of nature, from famine and
death . . . the production of the higher animals directly
follows.'[27] Ignoring the point, already made, that no such
thing follows on his theory, it is clear that Darwin, influenced
by Malthus, sees the natural world as a cruel competitive
battle. 'We may console ourselves', he writes, in a feeble
attempt at mitigation, 'with the full belief, that the war of
nature is not incessant, that no fear is felt, that death is
generally prompt, and that the vigorous, the healthy and the
happy survive and multiply.'[28]

One cannot deny that there is much suffering in nature,
that whole species become extinct, that death is the universal
fate of organic beings and that all animal life-forms have to
destroy other organic forms in order to survive. But does it
follow from this that it is appropriate to see life on earth as a
'war of nature', where the strong survive and the weak die?
Such a way of seeing the world is in stark contrast to a
theistic vision of the world, for which the earth belongs to
God, while humans are to hold it in trust, and for which the
righteous are to inherit the earth.

It is also in strong contrast with views that see nature as
a much more holistic, integrated web of interactions. Darwin
himself is not wholly blind to such aspects of the natural

world. 'How infinitely complex and close-fitting are the mutual relations of all organic beings,' he writes.[29] At this point, he seems to glimpse a very different vision of nature as a web of relationships, adapted beautifully to one another, in a developing and ever more complex harmony. Moreover, he occasionally seems to see this process as destined to continue in ways that increase beauty, harmony and complexity without limit: 'I can see no limit to the amount of change, to the beauty and infinite complexity of the coadaptations between all organic beings.'[30]

The metaphor of a war of nature here gives way to a different metaphor; that of a developing emergent whole, with increasingly complex and beautiful co-adaptedness among organic life-forms, and which pictures nature as expressing a continuous growth in harmonious complexity. Instead of selfish genes, ruthlessly competing, Darwin sees a finely balanced interplay of forms within an emergent totality, generating new states of organisation and beauty. This change of metaphor reflects a change that has also taken place in physics, from the atomism of Newton to the interconnected fields of energy that characterise relativity theory. In biology, the model of isolated units in competition has for some time now been opposed by a model of a unified web of interrelated and intricately balanced forces.[31]

On the newer, more holistic, picture, suffering and death are inevitable parts of a development that involves improvement through conflict and generation of the new. But suffering and death are not the predominating features of nature. They are rather necessary consequences or conditions of a process of emergent harmonisation which inevitably discards the old as it moves on to the new.

Competition and struggle exist, as parts of the mechanism by which organic life evolves to new and superior forms. But co-operation and self-sacrifice also exist, even at quite basic levels of conscious animal life. It is not just a blind will to power that drives evolution forward. It is also a striving to realise values of beauty, understanding and conscious relationship. This becomes quite overt at the level of *Homo sapiens*, but it is also plausible to see it as underlying the whole evolutionary process from the start. For it is, after all, that process which has produced *Homo sapiens*. The theist does not have to introduce any mysterious 'vital force' into the process, or any supposition that simple organisms are somehow trying to achieve higher goals by some semi-conscious effort. The theistic hypothesis is that there is a cosmic intelligence, God, which intends the evolutionary process to produce such values, and causes it to do so.

The goal of the process is a fully conscious goal, formulated in the mind of God. What God wills, and consequently what the process will eventually produce, is not the triumph of the strong, but the triumph of virtue, of beneficence, compassion and love. The ultimate evolutionary victory, on the theistic hypothesis, does not go to the most ruthless exterminators and most fecund replicators. It will go to beings who learn to co-operate in creating and contemplating values of many different sorts, to care for their environment and shape it to greater perfection. It will go to creatures who can found cultures in which scientific understanding, artistic achievement, and religious celebration of being can flourish.

EVOLUTION AND THE FALL

From a theistic point of view, the natural world that Darwin sees is a corruption of what the world is meant to be. It is not true that the will to power alone characterises the animal world. Rather, there is a quite extraordinary development of organised bodies and nervous systems which allow brains to form, which carry the first stirrings of conscious life. The natural world is a world of great beauty, and it proves to be conducive to the emergence of forms of consciousness that can react to and appreciate that beauty, which can rejoice in life, in its times of struggle as well as in its times of peaceful relaxation.

When one comes to the complex personal consciousness of humans, the picture is changed by the emergence of moral responsibility, of the possibility of self-centred action and of a willed self-alienation from the harmonious web of emergent organic relationships. The will to power, to dominate and use nature for selfish ends, only really comes to exist with the first human lives. When it comes to exist, it corrupts the natural course of things, as the human will to power corrodes and undermines the rhythms of the natural order. Far from denying that such a corruption exists, it is an essential part of the theistic vision that human life is estranged from its creator and is trapped in the bondage of self-will. But theists see this as a responsible choice that the first humans made, when they came to be able to make such a choice.

This entails a modification of early religious pictures of life on earth, in the biblical traditions, as having been created without suffering and death, which only came into the world after the selfish disobedience of Adam. Certainly we now know that suffering and death existed among animals long before the appearance of the first human beings. Bacteria and carnivores existed before *Homo sapiens* came into existence. Thus we must understand the death that the first conscious sin brought, not as a physical end of life, but as a spiritual death, a separation from God, the only true source of life. Sin, it seems clear, makes suffering, for all creatures encountered by humans as well as for humans themselves, much worse than it might otherwise have been. It creates a fear of death, now seen not as a natural process but as a possibly final separation from God.

The will to power and self is not, for the theist, the normal condition of all organic life, as Darwin's talk of a 'battle of nature' suggests. It is precisely a perversion of the natural tendency towards the realisation of value, a willed and conscious turning aside from one's natural, proper, destiny. The emergent web of nature is torn by human beings, as they turn aside from their divinely willed vocation towards lives of selfish desire. It is because humans are trapped in an alienated consciousness that they tend to project their own fears and passions onto the natural order itself, and see it as a blind struggle of selfish entities. They begin to see the basic carriers of heredity, the genes, as themselves 'selfish', the mutations that carry forward the emergent process as 'mistakes in copying', and the striving for new forms of life as 'struggles for survival'. These metaphors are all projections of a wholly negative and reductionist view of human existence onto the natural order. The theist thinks instead of genes as organic parts of the building up of complex carriers for consciousness, of mutations as vehicles of emergence, and of the striving for life as a striving for the realisation of new values. This is a much more positive view, and of course it derives directly from the hypothesis of God, as one who gives a positive goal and value to the natural order.

If one looks at the natural world, and asks which hypothesis seems more adequate to the facts, it seems to me that the 'emergent web' hypothesis is the more adequate of the two. The 'selfish gene' theory, which we shall be considering soon, has great difficulty in accounting for the genesis of culture, of scientific understanding and the powerful sense of moral obligation. It has to see these things as by-products of a struggle for dominance on the planet, and thus as without intrinsic value. It is extraordinary that they have come to exist at all, and no very convincing reasons have been given for their survival value. Moral obligations, in particular, often restrain conduct which would effectively exterminate one's competitors in the evolutionary battle, and scientists often show a biologically inexplicable preference for truth over expediency. The effect of the selfish gene theory is to undermine truth and moral obligations altogether, by claiming them to be only survival mechanisms, now perhaps counter-productive.

The 'emergent web' theory, on the other hand, gives truth and beauty a primary place in its account of the nature of things, and has no problem in accounting for the genesis of societies of rational moral agents, responding to moral and scientific imperatives in responsible freedom. It can also account for the 'struggle and death' aspects of evolution, as necessary conditions for the gradual emergence of ever more complex forms of sentient life. And it can account for the 'will to power' aspect of life on earth, as one that is introduced to the planet by the acts of responsible human agents who have chosen the path of egoistic desire rather than of dispassionate care. It gives a better account of all the relevant facts, and is thus the preferable hypothesis.

EVOLUTION AND PURPOSE

However, as with all such rather metaphor-based visions of nature, which one seems more appealing will largely depend on one's own personal reactions to existence and its value. The theistic hypothesis leads us to expect that this would be so, since a rejection of God will cause one to evaluate and react to existence in a much more negative and pessimistic way. Since people will probably continue to accept or reject God on very personal grounds, this means that there are likely to remain disputes in this area, based on deep personal evaluations rather than on strictly evidential considerations. If one asks whether the strong will inherit the earth, an answer will depend as much on one's own hopes and commitments as on an allegedly dispassionate survey of the evidence. For a theistic view, the 'strong' – those who depend solely on violence and oppression – will in the long run destroy themselves. The 'weak' – those who are prepared to give their

lives in the cause of love – will in the long run be supported
by the divine intention for the flourishing of goodness. The
theist hopes for the fulfilment of the natural order that
humans can bring about through a conscious relationship of
knowledge and love with the creator. It is the righteous who
will inherit the earth, because they are taken up into the
current of divine concern for the fulfilment of all things which
is cosmically irresistible.

I suggest that Darwin's theory of natural selection is
partly motivated by a gloomy and pessimistic view of nature
as a battleground of irreconcilably hostile forces. Malthus'
gloomy picture of human life seems to many contemporary
commentators much too atomistic and adversarial. Darwin's
gloomy picture of life on earth shares a similar worldview of
adversarial atomism. The *laissez-faire* capitalist ethic of
ruthlessly competing individuals lies just under the surface,
determining Darwin's choice of metaphors and images. Those
who find Lovelock's Gaia hypothesis more apt will see all
planetary life interlinked in a web of mutually supporting
relationships. This may help one to see the beauty and
wisdom of the natural world in a much more positive light.

Connected with his gloomy assessment of life is
Darwin's increasing inability to find any sign of purpose in
the evolutionary process. In my view, this is largely due to the
fact that he saw purpose very much in terms of particular
design. Any sign of randomness or malfunction in nature
would then count against the existence of a benevolent
designer. The thought that millions of species have become
extinct can lead to an impression of waste and destructiveness
in evolution which looks inconsistent with good design.

The best way to deal with this difficulty is to discard all
naive ideas of God as a parent who would like to eliminate all
waste and randomness if he could. Such ideas can stand in the
way of seeing the true purposiveness of the evolutionary
process. One must look at the evolutionary process in terms

of the underlying physical laws that drive it. These laws, far from being wasteful and random, are supremely elegant and efficient. It may seem that the element of randomness in genetic change is not very efficient. But the apparently random element is in fact the best way of achieving a goal-directed outcome, while leaving the process itself non-deterministic. Thus, a space is left for the free actions of intelligent beings, which will later be so important to the development of the cosmos. The apparently wasteful extermination of individuals and species is, in fact, the best way of achieving a gradual improvement of organic life-forms, while not permitting autocratic 'interferences' which miraculously effect improvements from outside the system. Moreover, it seems virtually undeniable that the process brings into existence states of very great value (like the appreciation of beauty, moral action and rational understanding), which could not otherwise exist in the same way. Thus the process is purposive, in the important sense that it is an elegant and efficient law-like system for realising states of great value.

I have suggested that natural selection alone does not provide a very good explanation of this fact. It makes the whole process highly improbable, is unable to predict what is likely to happen, and gives no reason for expecting any trend towards complexity and consciousness. By far the best hypothesis is that there is a cosmic mind of immense wisdom, creating a system which will shape itself to realise states of value. In that case, the existence of mind will not be an accidental by-product of a blind conflict of hostile atoms. It will be the fundamental reality that underlies the whole cosmic process. The intended goal of the process will be to produce minds, capable of creating and appreciating values. Natural selection is undoubtedly an important part of evolution. But to say that it wholly explains evolution is something that Darwin himself did not believe, and it is to fail to see the purposes and values that evolution alone is capable of realising.

NOTES

[1] Charles Darwin, *The Origin of Species by Means of Natural Selection*, p. 115.
[2] Ibid., p. 263.
[3] Augustine, *De Genesi ad literam*, 4:26.
[4] Arthur Peacocke is a strong supporter of such a view; see *Theology for a Scientific Age*.
[5] Richard Dawkins shows how large mutations would vastly decrease the chances of survival; see *The Blind Watchmaker*, pp. 72f. Tiny mutations would make cumulative selection too slow. Wholly arbitrary, or truly random, ones would undermine cumulative selection altogether.
[6] W. McCoy, *Journal of Theoretical Biology* 68, p. 457.
[7] Peacocke, *Theology for a Scientific Age*, p. 66.
[8] Ibid., p. 67.
[9] Gould's interpretation of the fauna found in the Canadian Burgess shale is that very different evolutionary paths might very easily have been taken, if just a few factors had varied: Stephen Gould, *Wonderful Life: The Burgess Shale and the Nature of History*, pp. 49f.
[10] A. Peacocke, *Theology for a Scientific Age*, p. 220.
[11] Ibid., p. 156.
[12] Darwin, *Origin of Species*, p. 215.
[13] Ibid., p. 69: 'I am convinced that Natural Selection has been the main but not exclusive means of modification.'
[14] Dawkins, *The Blind Watchmaker*, p. xiii: 'Our own existence once presented the greatest of all mysteries, but . . . it is a mystery no longer because it is solved.'
[15] Darwin, *Origin of Species*, p. 348.
[16] Ibid., p. 416.
[17] Ibid., p. 459.
[18] Ibid., p. 443.
[19] Ibid., p. 231.
[20] Ibid., p. 127.
[21] N. Eldredge and S. J. Gould, 'Punctuated Equilibria: An Alternative to Phyletic Gradualism'.
[22] This is the view taken by Arthur Peacocke in *Theology for a Scientific Age*.
[23] On this point I agree with Richard Dawkins, who writes: 'Mutation is not systematically biased in the direction of adaptive improvement, and no mechanism is known (to put the point mildly) that could guide

mutation in directions that are non-random': *The Blind Watchmaker*, p. 312.

[24] Peacocke, *Theology for a Scientific Age*, pp. 157–60. For further exposition of the idea, see Donald T. Campbell, 'Downward Causation in Hierarchically Organised Systems'.

[25] A. N. Whitehead, *Process and Reality*, p. 346.

[26] John Houghton, 'New Ideas of Chaos in Physics', p. 49.

[27] Darwin, *Origin of Species*, p. 459.

[28] Ibid., p. 129.

[29] Ibid., p. 130.

[30] Ibid., p. 153.

[31] A more holistic and positive picture of the biosphere is given by James Lovelock's Gaia hypothesis, outlined in *A New Look at Life on Earth*.

CHAPTER FIVE

The Metaphysics of Theism

IS GOD A SCIENTIFIC HYPOTHESIS?

Despite the strength of a theistic interpretation of evolution, a number of biologists continue to attack belief in God as somehow superfluous to an understanding of the natural world, and even see it as some sort of irrational prejudice. None writes more provocatively than Richard Dawkins. In a number of beautifully crafted books, Dr Dawkins takes Darwin's principle of natural selection, and strengthens and extends it to explain all the phenomena of organic and sentient life. In the course of this ambitious project, he takes every opportunity to deride theism, which he sees as a competing scientific theory to Darwinism. I intend to show that his understanding of theism is defective, and that he fails to distinguish clearly between scientific and metaphysical (or philosophical) beliefs. This leads him to misunderstand the nature of the conflict between science and religion, and to invent conflicts where none exist.[1]

It seems that Dawkins completely, even wilfully, misunderstands the nature of belief in God, and the sorts of reasons for which an intelligent person would assent to such a belief. The educated theist sees God as a self-existent being of supreme perfection, the source of all other beings which are generated for the sake of their goodness. This is not a scientific hypothesis. It is not a theory invented to explain particular occurrences in the world. What, then, is the idea of God for? God is primarily the supreme object of worship and prayer. God is known by the believer as a presence and power in and through all finite things, is apprehended in prayer as a being of unlimited wisdom, bliss and compassion, and is worshipped as the supremely perfect being of whom all finite perfections are images and reflections.

Believers do not infer God as an absentee 'first cause', or construct God as a speculative theory. They seek to know and love a reality of supreme perfection. They learn to do so in communities founded on the testimony of sages, saints and mystics who claim to have known that reality, and they are often encouraged to do so by some glimpses of such experiences in their own lives. The life of faith is a life of trust in the testimony of those whom one admires, of commitment to worship and self-transformation, and of loyalty to the deepest experiences one has had of transcendent reality and value.

Believing in God is a commitment to a self-transforming way of living in the world, in response to an experience of transcendent power and value, whether one's own or another's, received in trust. It involves one's deepest personal commitments, and orders all one's life towards a hoped-for vision of and union with God. It is not surprising, then, that for the believer, God is not a tentative hypothesis which one should always be seeking to test to destruction by actively searching for counter-evidence. That is rather like saying that a good marriage is best achieved by always seeking evidence of infidelity.

There are, of course, facts that may count against one's idea of God, and they must not be ignored. Yet it is entirely rational for a believer to leave such matters to others, and seek progress in the life of prayer as a basic commitment – even though it is one that may have to be renounced if arguments against it prove overwhelming. If faith fails, one comes to see that one cannot continue the path of self-transformation in seeking supreme power and value – and this could be because the world simply no longer seems to support such a course as appropriate. On the other hand, one may find that an increasing capacity for self-transformation, an increasing awareness of a guiding divine presence, and a sense of purpose in one's own life, strengthen belief in God.

God is a hypothesis in this sense, that if one commits oneself to a life of worship, this entails the belief that there exists a worthy object of worship. For theists, the truth of that belief entails that there exists a supremely perfect creator, since that is the only proper object of unlimited devotion. That in turn entails that any created universe will have a specific character – it will be intelligible, morally ordered and goal directed. Consequently, a demonstration that this universe is not rationally ordered, or that it is non-purposive or morally cruel or even indifferent, will undermine belief in God.

It is clear, then, that theism is falsifiable. Some (including Dawkins) think it has already been falsified. It is also confirmable, if the universe, as experienced, mediates, at least in part, a personal presence; if it is rationally ordered; if it seems purposive; if it seems conducive to the realisation of beauty and virtue, understanding and creativity; and if the idea of God seems coherent and plausible. Such confirmation cannot prove the existence of God, as a compelling inference for any neutral observer. For the atheist (who lacks any experience of God or rejects the values of worship and prayer), it nevertheless presents rational considerations for taking theistic claims seriously. For the theist (who founds belief on personal experience and a commitment to a life of prayer), it sets out considerations that place religious belief within a wider set of general beliefs about the world, and shows how it integrates coherently with them. This is a highly desirable, if not absolutely essential, aspect of belief in God.

GOD AND METAPHYSICS

As a matter of fact, the theist would claim that God is a very elegant, economical and fruitful explanation for the existence of the universe. It is economical because it attributes the existence and nature of absolutely everything in the universe to just one being, an ultimate cause which assigns a reason for the existence of everything, including itself. It is elegant because from one key idea – the idea of the most perfect possible being – the whole nature of God and the existence of the universe can be intelligibly explicated. It is fruitful because it is the basis of human confidence in the basic intelligibility of nature (so it is the basis of science), and it is the basis of confidence in the objective value and eventual triumph of truth, beauty and goodness (and so it is the basis of morality and of the affirmation of human worth and the meaningfulness of existence).

The level at which theism provides an explanation is the very general level of metaphysics. As such, it competes with other metaphysical theories like materialism. Materialism says that the only things that exist are material things in space. There is no purpose or meaning in the universe. Scientific principles are the only proper forms of explanation.[2]

These three claims are not scientific theories or assertions. They do not belong to physics or chemistry or psychology or biology. They are certainly statements of faith. Faith is not blind acceptance, as Dawkins wrongly claims, when he says: 'Faith . . . means blind trust, in the absence of evidence.'[3] On the contrary, faith is a basic commitment to a set of most general beliefs about the nature of reality, about what really exists. The faith of materialism is that there are no non-material or spiritual beings; that there is no purpose in the existence of the universe; and that there are no useful non-scientific sorts of explanation. If one examines Dawkins' language, one finds that these denials are supported by systematic mockery and demonising of competing views, which are always presented in the most naive light. They are expressed in highly emotive language. And it is implied that all intelligent or 'scientifically educated' people must agree with them. In other words, at these points, Dawkins depends on propaganda and rhetoric. That is, of course, because the materialist theses are non-provable and highly disputed among intelligent people, and Dawkins knows it.

Materialist beliefs are not parts of science or implied by good scientific practice. Newton's laws are true whether materialism is true or false (as a firm believer in God, Newton thought it was false). You could say materialist beliefs are superfluous, and should be ignored. But the fact is that the way we live our lives often assumes a belief about them, one way or another. They are in a sense explanatory – they say what the world is like, and why it behaves as it does. But they do not explain any particular events. There is evidence for and against them. That is, there are facts that count for or against them, though none of these facts are conclusive. Yet materialist beliefs are not *based on* evidence, in the way Newton's laws are based on repeated experiments. They are beliefs which tell you what you should be prepared to count as evidence. They set out a basic worldview.

Dawkins shows very well how one can be completely, passionately devoted to them, even when one realises that they are not completely provable or universally accepted. One holds them because they seem to form the basis for a coherent, adequate and consistent description of the world which fits one's fundamental value-judgements and attitudes. But the fact is that there is more than one claimant to such a metaphysical scheme. In all these respects, materialism functions just like theism, as one competing metaphysical scheme amongst others.

Theism does not compete with science, but it does compete with materialism. Dawkins' systematic confusion of the two is not helpful – but then it is not meant to be. This is because one of the main propaganda devices Dawkins uses is to persuade the unwary reader that 'science' (which he assumes is obviously true) is incompatible with theism. To do that, he has to confuse science with metaphysics, and deceive the reader into thinking that all respectable scientists are really materialists – which is as false a belief as most that one can think of.

Theism is, then, a metaphysical hypothesis. It is incompatible with materialism. But it is not incompatible with science. The root of materialism is probably a firm commitment to empirical scientific method as the only reliable way to discover truth. Commitment to experimental method is in itself entirely commendable. But when it begins to exclude every other understanding of truth, one may suspect that it will result in a radically impoverished view of reality as a whole. The root of theism is probably a commitment to worship and prayer, which carries with it the belief that aesthetic, ethical, personal and relational aspects of experience provide distinctive paths to truth, and that the highest truth of all lies in apprehension of an objective reality of supreme beauty and goodness.

Is God Superfluous?

Both materialism and theism express very definite value-commitments, and that is the real source of their emotional power. They might be seen as basic evaluative perspectives, which attempt to organise and explain all aspects of human knowledge and experience from their own viewpoint. The 'truest' perspective will be the one that includes, systematises and relates the widest range of data in the most economical and elegant way, without distortion. Where the collection of 'evidence' comes into this process is in the provision of the widest possible set of data, and its arrangement in a comprehensible way within a purportedly comprehensive scheme. Materialism scores well on economy and elegance. But it scores extremely badly on comprehensiveness, ignoring completely all those features of personal conscious experience

and purpose with which we are in fact most familiar. Theism
has its drawbacks, which arise from human beings trying to
see how the universe might look from God's point of view.
But it scores very well on comprehensiveness, and is
economical and elegant. It looks to me as though theism,
though not existing in a wholly adequate formulation (which
is hardly surprising, given the limitations of human intellect),
is the most adequate metaphysical hypothesis there is. It is
important to keep in mind, however, that it is not based on
abstract speculation, but on a practical commitment to a life
of worship and prayer. Moreover, it is in practice not a mere
theory, a piece of armchair theorising, but a deeply felt vision
of the nature of human existence.

What is this vision? It is that the whole universe, from
the furthest star to the air we breathe moment by moment, is
held in being by a reality of unlimited power, consciousness
and goodness, by infinite spirit. This reality is beyond time
and space and every limitation, and therefore beyond human
thought itself. Yet all things are in some ways expressions of
its reality, and it can be sensed in and through the goodness
and beauty of creation. 'The heavens are telling the glory of
God, and the firmament proclaims his handiwork.' (Psalm
19.1) Many people, perhaps most, occasionally experience a
sense of something transcendent, something beyond decay
and imperfection, yet somehow mediated by special places or
circumstances. These epiphanies of Spirit seem to speak of a
reality 'far more deeply interfused' (Wordsworth, *Lines
Composed a Few Miles Above Tintern Abbey*), of 'Infinity
held in the palm of a hand, eternity in an hour' (Blake,
Auguries of Innocence). It is in such moments that one may
have a sense of the dependence of all finite and conditioned
things on an infinite, unconditioned reality of pure Spirit,
which can be known within, by silent waiting or by simply
attending to glimpses of beauty and wisdom in the world.
Perhaps religious faith begins, for most of us, in such small

epiphanies, in 'a sense and taste for the Infinite'. It is from such glimpses of a spiritual reality underlying this phenomenal world that one may develop the desire to seek a deeper awareness of it, and, if possible, seek to mediate its reality in the world. If that happens, religious faith is born. Worship and prayer are, basically, ways of deepening this awareness and transforming the self to reflect and mediate the divine Spirit. Only at a much later stage may one feel the necessity to develop a coherent metaphysical theory, arising out of and expressing this primal vision. It is only then, at that stage of reflection, that the theist must engage in argument with the materialist.

The minimal theistic claim is that science is not incompatible with the existence of God. Dawkins sometimes accepts that this is true, at least for sophisticated theists. He then only argues that theism is superfluous. It should be obvious, however, that he is here simply missing the point of theism. God might be superfluous as far as physics or chemistry go. That is, there is no need to mention God in doing physics, and we would not introduce God to explain why some physical phenomenon occurs. That is not really very shocking. I do not need to introduce God to explain why the car engine starts when I turn the ignition key. God is superfluous to car mechanics. But that does not mean that God is completely superfluous. God actually provides a better explanation of the facts uncovered by science than materialism does. This is still not a scientific explanation. It is a metaphysical one. But the theistic hypothesis is a better one than the materialist hypothesis, because it makes the existence of a universe like this very much more probable than materialism does. Although the believer's acceptance of theism is not *based on* its superior explanatory power, faith is shown to be both rational and illuminating if one can see that it is not a blind acceptance of the absurd, but a deeply satisfying explanation of how and why the universe exists.

One must keep in mind that the concept of God is not primarily an explanatory hypothesis at all. Its importance for the believer lies in the fact that it is essential to the rational practice of worship and prayer. The theist sees the world as mediating the presence of personal Spirit, in a way not wholly dissimilar from the way in which certain material bodies mediate thoughts, feeling and intentions. Just as one can investigate the physics and chemistry of the human body, without considering anything about personal and moral qualities, so one can investigate the physics and chemistry of the physical world, without considering whether or not it mediates a spiritual reality and power. But the human personality of a particular body is not superfluous when we fall in love with someone whose body it is. So God is not superfluous when we apprehend the beauty and power of supreme Spirit. God is not a scientific theory. God is a personal reality of supreme perfection, to whom persons can be related in knowledge and love. That is the basis of belief in God, but it entails that God will also be explanatory of the nature of the universe, in a strong metaphysical sense.

NOTES

[1] In this, I am intending to supplement the admirable study of Dawkins' thought by Michael Poole, in *Science and Christian Belief* 6, pp. 41–59, continued, with a response by Dawkins, in vol. 7. Another very good discussion of the issues is David Lack, *Evolutionary Theory and Christian Belief.*

[2] There are not many real materialists among philosophers. One sophisticated account is given in Anthony Quinton, *The Nature of Things.* Quinton would not accept the third thesis mentioned, and it is a very extreme view. However, Dawkins seems to accept it.

[3] Richard Dawkins, *The Selfish Gene*, p. 212.

CHAPTER SIX

The Elegance of the Life-Plan

THE CUMULATIVE PROCESS ARGUMENT: THE INCREDIBLE SHUFFLE

There are seven crucial stages in the evolution of life on earth, each of which requires a precise set of coexisting properties for its existence, and which cumulatively build to form the world as we see it today. These stages correspond quite closely to the series of 'thresholds' which Dawkins outlines at the end of his book *River out of Eden*.[1] But whereas he sees these thresholds as just occurring by chance – that is, without any design or 'weighted' directionality – I shall try to bring out the way in which such a hypothesis is implausible for each stage taken separately, and vastly implausible for the whole series. The whole process is just too well structured and integrated to be plausibly seen as an accident, if there is any better explanation available. And of course there is.

The first vital and necessary stage in the formation of life in the universe is the genesis out of an earlier mass of superheated energy of simple interacting nuclear particles, which can unite to form stable complexes. Quarks unite to form protons, neutrons and electrons, which in turn unite to form atoms. Atoms are so shaped that they combine in incredibly precise chains. For example, the haemoglobin of our blood consists of four chains of atoms, linked in 574 units, in a completely precise and unique way. Without such a linkage, the subsequent development of organic life would be impossible. 'There is no need to think of design', writes Dawkins, as one would expect.[2] For him, the superhot energy of the Big Bang just happened to cool into particulate molecular shapes which could build stable complexes of this intricate form. Such a thing is indeed vastly improbable, given all the ways, perhaps an infinite number, in which energy could have cooled and degraded, without forming any stable complex unities at all. But then, Dawkins says, anything is improbable, and in an infinite time, this improbable thing may just happen to come about.

There are two arguments here which, on closer examination, turn out to be fatally weak. One of Dawkins' main arguments is the 'cumulative process' argument. It is unbelievable that a very complex form could just come into existence out of nothing. But, he says, if one can account for it as the result of thousands, or millions, of very small developments, each one of which is only slightly improbable, then we have achieved as good an explanation as could be desired. So all we have to do is to start with a set of very simple elements and push them about in various ways. First, two or three might stick together; then larger groups may form; and at last, we have haemoglobin strings, of astonishing complication, but built entirely out of many repetitions of simpler combinations, stuck together.

Think of dealing a pack of cards. If they came out perfectly ranked in suits, we would be astonished. But suppose you have an unlimited number of deals. Each time, the cards are not completely reshuffled, but cards dealt of the same colour or of adjacent numbers are retained in that order. We would not be very surprised at getting pairings of this sort. If we keep repeating the process, eventually we will get all cards of the same colour together, and even later we will have achieved a deal of perfect suits.

At first this trick seems to reduce our surprise. But consider further. What surprised us at first was that no one had planned the deal, yet the cards emerged in a way highly significant to us. They formed an exact order a designer might have planned. The order in itself is of no significance – it is just a matter of marks and colours on pieces of card. What is significant is that we give meaning to those marks and colours, such that one ordering of them is of high value, is highly desirable, and most others are not. It is only when we have assigned a high value to one ordering of the cards that the deal of such an order is astonishing. It is just what we want, yet no one planned it. It could happen by chance, yet it is such an unlikely occurrence, and it so exactly fits our desires, that we would not believe it was chance.

Now we set out to reduce our astonishment by setting up a rule that will ensure that the result we desire will be accomplished after a good deal of time and effort. We must have a random shuffling of a precise number of cards, with certain combinations retained each time, until the desired sequence emerges inevitably, though unpredictably in detail. Obviously, we will not be surprised when the sequence at last emerges, because we have now designed the system so that it is bound to do so. Yet the final result is still what we desire. And is it not highly unlikely that there should be a rule which ensures that what we desire will come about? It is, if anything, even more unlikely that there should be such a rule than that our first deal should produce the sequence we want. For we can calculate exactly the probability of one sequence of fifty-two cards being dealt, by counting the cards and their possible combinations. But we cannot calculate exactly the probability of there being a rule that will inevitably produce the sequence we want, compared with all other possible rules for dealing cards, of which there must be an indeterminately large number. With a bit of effort we could work out such a rule. With a bit more effort, we could work out the most efficient rule to produce the result we want. But it would be truly amazing if the most efficient rule for the job already existed, just by chance.

Dawkins' strategy for reducing amazement and incredibility does not work. It just shifts the surprise from the spontaneous generation of a complex and highly desired result to the spontaneous existence of an efficient rule which is bound to produce the desired result in time.

THE COSMIC CARROT: OR WHY THE SIMPLE IS NOT MORE LIKELY

The amazing thing is that the basic 'rules', or laws of physics, are such as to produce atoms capable of combining into molecules and assembling themselves into the fantastically complicated strings needed to produce organic life-forms. Yet the laws of physics themselves emerged from a primal Big Bang, when there were no nuclear elements for the laws to operate on. So where did those laws come from? How did they come to take the supremely efficient forms they did take? And how did they come to interrelate with one another so as to make possible a coherent, intelligible universe? As the physicist Eugene Wigner says, 'The miracle of the appropriateness of the language of mathematics for the formulation of the laws of physics is a wonderful gift that we neither understand nor deserve.'[3]

Ah well, says Dawkins, there is no answer to those ultimate questions. But at least the basic elements and laws are rather simple, so it is more likely that they just popped into existence for no particular reason than that they were designed by a cosmic mind of supreme wisdom. Any mind capable of designing such an intricately structured universe as this would have to be extremely complex. So, it is actually more likely that the relatively simple laws of nature would come into existence for no reason than that they would be designed by a God whose being would have to be more complex than they are. It is thus more probable, overall, that there is no God.

This, however, is the second rather weak argument, the simplicity argument. Suppose the basic laws of physics popped into existence for no reason at all. One day, they did not exist. The next day, there they were, governing the behaviour of electrons and atoms. Now if anything at all might pop into existence for no reason, there is actually no way of assessing the probability of laws of physics doing so. One day, there might be nothing. The next day, there might be a very large carrot. Nothing else in existence whatsoever, but there, all alone and larger than life, a huge carrot. If anything is possible, that certainly is. The day after that, the carrot might disappear and be replaced by a purple spotted gorilla. Why not? We are in a universe, or a non-universe, where anything or nothing might happen, for no reason. Why does this thought seem odd, or even ridiculous, whereas the thought that some law of physics might just pop into existence does not? Logically, they are exactly on a par.

At least, Dawkins suggests, it is more likely that simple things should pop into existence than that complex things should do so. This can be shown to be a false assumption, however. It is the 'fallacy of simplicity', the fallacy of thinking that, where possibilities are unlimited, it is more likely that simple possibilities should become actual than that complex possibilities should. One can see the fallacy, if one reflects that there are obviously a great many more complex possible things than there are simple possible things, since one can always combine simple things in many different ways to make complex things. Four simple things, for example, can be combined in different ways to make twenty-four different complex things. There are always many more complex possible things than the simple possible things of which they are composed. So it is, if anything, more likely that a complex thing might jump into existence than that a simple thing might, since there are hugely more of them, waiting to jump.

It is therefore not less likely that a complex being like God should pop into existence than that a set of laws of physics should do so. There is no greater probability of laws existing without God than of God existing as the creator of the laws of nature. Of course, there is something wrong with the thought that God is the sort of very complex being Dawkins postulates, and also with the thought that God might 'pop into existence'. But suppose for a moment that it makes sense. The point is that the existence of God is not a priori less likely than the existence of laws of physics. Now it has just been seen that the laws of physics are efficiently ordered so as to produce highly desirable states. They thus show every sign of being designed, and God is just the being to do the designing. So it is very much more likely that the laws are designed by God than that they just come into existence, or just happen to exist. Like the rules for producing desired sequences of cards, it is likely that some mind has thought them up.

Dawkins' objection to introducing God is that any mind capable of constructing a universe must be very complex, and so cannot constitute a truly simple explanation. There is a great danger of confusion, however, about the idea of 'simplicity'. I have already distinguished two senses of simplicity, the exclusive and the inclusive (pp. 84f.). The exclusive sense works by excluding everything complex, or by explaining the complex in terms of a combination of simpler elements. This sense has a perfectly proper use in science. Thus, the enormously complex behaviour of gases in a confined space can be very well explained by formulation of a fairly simple law, Boyle's law, relating pressure, temperature and volume. One has a small number of easily measurable variables and a small number of easily calculable laws. Here is a paradigm case of the explanation of complexity by appeal to a simple schema.

That there should be such a small number of variables and laws, and that they should be so easily measurable and calculable, is a wholly fortuitous fact about the world. Indeed, it is vastly more improbable that such simplicity and elegance should exist than that there should be unmeasurable chaos and incalculable and irreducible complexity in the cosmos. This is, in fact, Wigner's 'miracle', which we have no right to expect. Far from simplicity of this sort being more likely than complexity, it is very much more unlikely, and its existence is in itself suggestive of intelligent design.

In the case of Boyle's law, nothing of significance is being excluded by the provision of a simple explanation of complex behaviour. Such accounts become 'excluding' only when they are misused to suggest that, in every case, the complex is nothing more than the result of combining the simple elements isolated by some theory. For instance, it might be suggested that human thinking and intending are nothing more than electro-chemical activity in the brain, and that a neuro-physiological account of such activity would give a complete explanation of some thought process. It might well be thought that things would be simpler, in some sense, if this were true – one could stop talking about thoughts and intentions altogether, and thus eliminate a whole set of difficult concepts. One could never, in practice, complete such a translation; it would be much too time consuming and difficult. But a more severe problem is that it would simply exclude a lot of interesting data, pretending that it does not exist. That is when appeal to 'simplicity' becomes misleading. Scientifically simple accounts of the universe are often very helpful, but if they are taken to provide the whole truth about everything that exists (which no responsible scientist would ever claim), they are probably going to be immensely misleading.

It should be obvious that God is neither simple nor complex in these senses. God is not complex in being an unmeasurable chaos. God is not simple in consisting of just a few basic properties and algorithmic laws relating them. But there is another sense of simplicity, which I have called 'inclusive simplicity'. It might equally well be called 'unitive' or 'integrative' simplicity, because it refers to the way in which some concept can integrate many different sorts of data in an illuminating way, or in which some being can unite many different sorts of elements in one organic and interrelated unity.

The concept of God is simple in this sense. It is the idea of just one basis of all possible finite beings, which originates all other realities for good reasons, and realises the highest compossible set of values in itself. It is thus the simplest possible and most all-inclusive integrating concept. If it is, as most theists have always claimed, self-explanatory, then it will answer the question 'Why does it exist?' in the most adequate possible way; by showing that it is of supreme value, and that it is wholly intelligible. God is not, as Dawkins supposes, a complex reality, as though God was rather like a human mind, made up of lots of disconnected thoughts, plans, desires and feelings. God is simple in a very special sense, as being the one self-explanatory and supremely integrating reality. God can explain why the laws of nature exist as they do – because they are supremely well designed to produce sentient beings by a process of evolution in a material universe. God is not just an arbitrary additional entity, but the purest sort of unity, which includes and unites all possible complexity within itself.

But it might still seem that one needs to answer the question with which the eleven-year-old Bertrand Russell tormented his nanny: who designed God? Or, to put it in the terms used so far, if it is very improbable that there should be efficiently ordered laws of physics, is it not just as improbable that there should be an all-wise God? What I have shown is that it is not very improbable that there are efficient laws of physics, if there is a God. So God is a good explanation for the laws of physics. But is not the existence of God itself very improbable?

WHY GOD IS NOT PROBABLE (BUT ABSOLUTELY CERTAIN!)

The answer is that it is very doubtful whether the idea of probability applies to God at all, logically speaking. Probability is a very difficult notion, much debated by

philosophers. I have so far regarded an improbable event as
one to which there are a great number of real alternatives,
with no cause or reason why one event should occur rather
than the others, or perhaps even with some reason, though
not an overwhelming one, why the 'improbable' event should
not occur. My winning the lottery is improbable, because
there are so many sequences of numbers other than mine that
could come up. A slightly stronger sense of improbability is
when there exists a law that ensures a specific outcome most
of the time, but allows exceptions. The exceptions are
improbable, in that there are alternatives which there is a
reason for expecting. We might say that a probable event is
one that happens in many or most possible worlds. An
improbable event is one that happens only in one or a few
possible worlds. There are other senses of probability, too,
some of them much more mathematical. But we might say
that an improbable event is always one to which there are
alternatives that we have some reason to expect to occur.

If we apply this to laws of physics, there is reason to
expect that, out of all conceivable states of affairs, any laws that
exist would be less efficient or coherent than they are, so they
are truly improbable. The reason is that there is a huge, perhaps
infinite, set of possible universes with laws different from those
we have. But very few of them are capable of producing the sort
of order and complexity that is necessary for the existence of
sentient life. Even given the fact that we cannot imagine what
very different life-forms might be possible, we can see that very
few sets of possible laws will be able to produce any coherent
forms at all. Most possible universes will never get going,
because they will not have the requisite stability and complexity
of form. Some physicists argue that this is the only universe that
could produce conscious life-forms, because of the complexity
and precise integration of the mathematical laws that underlie it.
Without going that far, it is clear that the number of unstable
and therefore non-viable possible universes vastly outnumbers

the number of viable possible universes. Therefore the existence of a universe as stable as this is highly improbable.

If there is a God, however, then the existence of such laws, if they are indeed efficiently ordered to produce a desirable outcome, far from being improbable, is rather probable. For a God might well invent such laws for a good reason. But what about God? Is there not reason to expect that God would be less perfect or wise than God is, or even that there will be no God? God will then be very improbable; and we know that Dawkins and his fellow believers think God is extremely improbable, perhaps even impossible.

Traditional theists do not, however, reply to this as one might expect, by saying that God is actually very probable. The dispute is not about whether God exists in one, or a few or in most possible worlds. It is about whether God exists in all possible worlds. If God did, then God would not be probable. God would be completely certain, even if atheists denied it until they were blue in the face. (Remember that I am using the expression 'possible world' here in a logical sense, to mean 'everything that exists, whether natural or supernatural'.)

It may seem that there is a very obvious proof that God does not exist in all possible worlds. All one has to do is imagine a universe with God, then take God away from it, and one has a possible world without God. However, it is not quite as easy as that. You might imagine a world without God, just as you might imagine yourself flying back through time, but can such a thing be? What we can imagine, or think we can imagine, is not a good guide to truth, as any mathematician can quickly prove! For instance, we may think we can easily imagine a square which is exactly equal in area to a given circle. But mathematicians can prove that such a thing is impossible. Our imagination is not always reliable. We cannot prove, just by imagining it, that a world can exist without God. But maybe we cannot prove there must be a

God either. So we seem to be stuck without a proof either way. To say that we cannot prove it, either way, is emphatically not to say that there are no reasons for or against believing it. There are many reasons on both sides, as it happens. But none of them is quite conclusive, and they strike different people with different force.

Still, one thing should be clear. God is either necessary or God is impossible. It is not a matter of probability. Theists say that God is self-existent, depending on nothing other than the divine being itself. God cannot fail to exist. God cannot come into being or pass away. God is the very source and root of all existence. God exists by necessity, and there is no alternative to the divine existence. Atheists, on the other hand, say that the idea of a being who exists in all possible worlds is incoherent, logically impossible (because, if it were possible, and since this world is a possible world, that being would exist in this world, and so it would actually exist). It seems that one cannot prove either that God is necessary or that God is impossible. Nevertheless, God is either necessary, and so God exists, or God is impossible, and so could not possibly exist. Either way, the existence of God is not more improbable than the existence of blind laws of nature. We are talking about a different, unique sort of reality.

It is actually very important to see this, if one is going to be fair to traditional theism (which Dawkins is not very concerned with, even though he professes a great love for truth and careful reflection in other areas). For theists in the great religious traditions of the world, God is not just a cosmic mind which may or may not exist, and which is like other minds, but bigger and better. God is a totally unique sort of reality, a cosmic mind indeed (though even that is a very inadequate idea of God), but one that exists in and from itself alone, without companions or equals, existing in and as the sole basis of every possible world.[4]

I conclude that both the 'cumulative process' argument

(see pp. 106f.) and the 'superior probability of the simple' argument (see pp. 108ff.) are unconvincing. The existence of rules that ensure that complex and relatively stable structures emerge in the universe suggests an intelligent rule-giver. On a theistic view, God is such a rule-giver, and the existence of God is either impossible or necessary. Thus the existence of God either has a probability of zero or of one (absolute certainty), even though we cannot know for certain which! If the latter, it is obvious that the existence of God is much more probable than the existence of very simple bits of matter, existing for no reason at all – even though, as I have pointed out, it can be misleading to speak of God's existence in terms of probability at all. With regard to the first stage of the life-series, then – the genesis of nuclear particles which can unite to form stable complexes – the hypothesis of God is preferable to that of a cumulative, but accidental, development from simple brute matter to highly organised complexity.

THE FALLACY OF COSMIC PROMISCUITY

The second crucial stage in the evolution of life is the formation of self-replicating molecules from complex combinations of chemical elements. Dawkins' view is that the existence of replicating life-forms is 'exceedingly improbable'.[5] Complex chains of molecules build up which have the extraordinary property of producing copies of themselves. The structure of DNA, with its intertwined molecule chains, is such that it builds up a copy of itself by attracting other molecules, forming them into a pattern and shedding them. There is no antecedent likelihood that this would ever happen. On any purely physical account of the shuffling and recombination of molecules, according to Dawkins himself, it is unlikely in the highest degree.

However, Dawkins suggests that, given enough time,

even very improbable things might happen, and that is that. He even supposes that 'given infinite time, anything is possible'.[6] This is a good example of the fallacy of cosmic promiscuity. This fallacy says that everything that can happen, will happen, given enough time. If there is an array of possible states, they will all come into existence at some time. Once you put it like that, it can be seen to be clearly false. There is no reason why even one possible state should ever come into existence. There is no reason why, if there are *n* possible states, they should all come into existence one after another, or at some time. Perhaps states get into a 'loop', and a small group of them keeps recurring, so that there are always some states in existence, but the vast majority of possible states never get actualised. If, as seems likely, there is an infinite number of possible states, they cannot all come into existence, since however many of them have existed, there will still remain an infinite number that have not existed. Moreover, as I have pointed out, some possible states (like the state with a creator God in it) exclude many other possible states (like the state with no creator in it). So it is logically impossible for all possible states to exist. That is why Dawkins' statement embodies a fallacy.

Dawkins is, of course, assuming that there are basic laws of physics, which will recombine nuclear particles in various more-or-less automatic ways, and in all their possible combinations. What is then meant by saying that the genesis of self-replicating chains of molecules is exceedingly improbable is that the degree of organised complexity required is so great that its generation, given only basic laws of physics, will be exceedingly rare and may well never occur (since it is not true that every possible state will occur some time). I have already suggested that the existence of such laws of physics, and of elements over which they operate to produce such complex forms, is itself hugely improbable. But, given their existence, the formation of self-replicators is, however unlikely, possible.

The theist cannot deny that DNA replication could have come about by an exceedingly improbable chance. But if this is the best the materialist hypothesis can do, it is vastly inferior to any hypothesis that makes the generation of self-replicators highly probable. The hypothesis of a God postulates a being who designs the basic laws and elements, and whose 'top-down' holistic influence conserves ('selects for') those molecular combinations that begin to build up replicating chains, so making the generation of such chains certain, given time. The hypothesis of God makes the existence of replicators much more likely than does the materialist hypothesis, and it is therefore to be preferred.

WHY MUTATIONS ARE NOT MISTAKES

Once replicators exist, highly complex life-forms obviously cannot exist unless there is some development from the original relatively simple forms. An elegant way to bring this about is to devise a system such that the process of replication does not remain forever unchanging. Rather, as replication occurs, there must be changes in structure which will form cumulative developments. For this to happen, the changes, or mutations, must be small enough to preserve the general structure of the organism, yet large enough to give rise to a significant change in either form or behaviour. Mutations must effect coherent and co-ordinated modifications in organisms, and they must overall tend to result in progressively more complex structure and organisation. All these constraints form the third crucial stage of the evolution of life on earth.

From a combination of elements, bound in the symmetry of the self-replicating double helix, forms of being capable of reproduction, movement and eventually of consciousness, originated. This progressive development

towards life-forms capable of consciousness and action is just
what one might expect if the process was devised by a being
that intended to develop conscious life from a simple material
base. For Dawkins, however, all mutations are 'mistakes' in
copying, and the process is seen as a blind reprocessing of
molecules, without goal or improvement.

The Darwinian sees mutation as random. That is, it
carries 'no general bias towards bodily improvement'.[7] It is
neutral or harmful most of the time, and it is an accident that
some mutations cause the production of more stable or
complex life-forms. All so-called 'improvement' comes from
the cumulative effect of natural selection. 'Natural selection is
the very opposite of random',[8] writes Dawkins, since it works
by the repeated application of one simple truism – that
organisms that replicate more efficiently will become more
numerous and survive. For the Darwinian, all improvement is
the result of natural selection, which is just to say that some
mutations will cause replicators to become more efficient, and
they will survive best. There is nothing in the mutating
process itself that leads to any improvement. At one point,
Dawkins dismisses the alternative view that mutational
change does tend towards increasing bodily organisation as
'mystical nonsense'.[9] More calmly, he says, 'nobody has ever
come close to suggesting any means by which this bias could
come about'.[10] 'No mechanism is known (to put the point
mildly) that could guide mutation in directions that are non-
random.'[11] The view that there is such a bias, which he calls
'mutationism', is so wrong, he says, that it could never have
been right, and is so absurd that it does not even need to be
disproved. So 'it is selection, and only selection, that directs
evolution in directions that are non-random with respect to
advantage'.[12]

There are two questions to raise here. Is mutation really
quite as random as Dawkins makes out? And is natural
selection the only factor to account for the observed apparent

improvement in biological adaptation and complexity? The first question sounds like a properly scientific one; and indeed Dawkins implies that it is a settled scientific question. Genes, sequences of DNA, do not mutate in any particular direction. At other times, however, he is more circumspect and admits that this is a hotly disputed issue in biology. I am content to leave that dispute to professional geneticists, since it is more to the point to concentrate on what is meant by 'random' change.

Random change is obviously not the sort of change in which anything at all might happen; it is not chaotic. One of the basic faith-postulates of science is that there is no ultimate chaos in nature. That is a postulate which theists would strongly support, since they see nature as the product of a wholly wise, intelligible God. Mutation does not occur in a vacuum, or for no reason. It consists in a recombination of the components of DNA, and it is caused by such things as X-rays, radioactive substances and various chemicals. Moreover, as Dawkins points out,[13] it is influenced by environmental factors such as its embryonic location and its relation to other recombining elements. This leads one to suspect that mutation is only random in that we cannot detect or specify the causes leading to genetic change.

When one takes the whole physical environment into account, it seems to be false that genetic changes carry no general bias to bodily improvement. For the most obvious thing about them is that they have led to continuous bodily improvement. As Dawkins points out in his discussion of the evolution of the eye, 'serviceable image-forming eyes have evolved between forty and sixty times, independently from scratch, in many different invertebrate groups'.[14] What can this suggest but that there is a seemingly inevitable trend in the physical order to continuous bodily improvement (assuming that the eye is an improvement, of course)?

'IT WAS RIGGED': THE EYE AND OTHER STORIES

When Dawkins talks about random copying mistakes, what he means is that DNA sequences are recombined, largely by the action of external physical influences, and that the vast majority of recombinations are neutral with regard to adaptive improvement (with regard to generating features which will, directly or indirectly, increase replicatory efficiency). Some recombinations will impair replicatory efficiency (by killing off the animal concerned), and some will increase such efficiency (for example, by increasing the acuity of the eye). In other words, purely genetic changes do not all, or even mostly, tend towards improvement. Nevertheless, it is clear that some changes generate improvements in replicatory efficiency, and it is these that will become most pervasive in time. The process as a whole is set up so that there is a bias towards bodily improvement, perhaps even an inevitable one.

The remarkable fact about all this is that a reshuffling of DNA sequences should produce any improvements at all, much less a continuous series of improvements. Dawkins, in writing about the development of the eye, cites the work of Nilsson and Pelger, who set up a computer simulation to calculate how long it would take to evolve an eye from a light-sensitive cell, by a succession of steps, each of which involves an improvement in visual acuity or spatial resolution.[15] They found that something like a human eye would evolve, by changes in the living tissue, in fewer than 400,000 generations (less than half a million years). Changes consisted in the tissue becoming larger or smaller, thicker or thinner, more or less refractive. Such changes were 'random', in the sense that all these possibilities were taken to occur in some individuals in each generation. The vital factor is that in each generation there is some small mutation which is an improvement, and which is conserved. Furthermore, the process of improvement is bound to continue, as long as the various physical properties vary in every physically possible way by small amounts in each generation.

What this suggests is that the basic physical properties have been carefully selected so that continuous small variations among them, caused by various environmental factors, would eventually produce a desired outcome – an eye sensitive to its environment. It is far from obvious that there will exist just that set of properties – transparency, thickness, size and light-bending capacity – that will lead to the existence of eyes, or that there will exist the sort of mutating influences that will cause continuous small variations. What is needed to get a developed eye is an initial organisation of related properties which will produce a light-sensitive lens in the first place. 'It would be nice', Dawkins says blithely, 'to do another computer model . . . to show how the first living photocell came into being by step-by-step modification of an earlier, more general-purpose cell.'[16] But any such model would itself have to begin with exactly that organised and correlated set of properties that could give rise to a photocell by the desired method. In other words, the initial state would have to be rigged, by very careful and precise selection, in order to produce the desired result. This does not suggest that the process is a blind, non-purposive one. Quite the contrary; it suggests a very high level of initial organisation.

After the initial setting up of the system, one would have to arrange for physical interactions with cosmic rays or other forms of mutating influence which would cause continuous but suitably small variations in the relevant properties. This again requires a very fine balance to ensure that the chosen forms of mutational causality do not simply destroy the genes from which photocells develop, or cause such gross mutations that no step-by-step form of development is possible. It looks as though the whole system of physical nature, with all its forms of radiation and other kinds of mutating influence, has to be rigged, as well as the initial structure of the photocell, to produce the desired result – something that would require huge intelligence and power.

Finally, some of the mutations have to be selected and preserved, if the eye is to be formed. In the computer model, this selection is done by the human operator, who decides that all improvements in visual acuity should be preserved as a base for subsequent modification. The operator has a definite goal, a final state to be achieved, and sets up the program so that it will achieve that goal. Dawkins assumes that natural selection will do this selecting job. But natural selection has no goal, and does not desire the existence of things like eyes at all. It only conserves those properties that increase the efficiency of an organism's reproduction.

Will a tiny increase in the visual acuity of a photocell (which is what one generation of mutations achieves) increase the reproductive efficiency of a whole organism to a large enough extent to ensure the spread of the more acute cell across a whole range of descendants? The story is that slightly better eyesight will enable animals to avoid death and find food slightly more effectively. So they may survive longer, and reproduce more than those without the improved sight. This is quite true, given the existence of some very important necessary conditions. The eye must be correctly connected to a brain, within a complex organism which is able to react to visual stimuli. Not just the eye itself, but millions of correlated cells, must continue to function and change in ways that do not cancel out the small visual advantage obtained by mutations. The environmental conditions must remain such that visual acuity is an advantage – light must be reflected from surfaces, gravitational pull must remain fairly constant, and no environmental disaster must occur that obviates any slight visual advantage – for example, by wiping out partners who make reproduction possible. Not just the body of the organism, but its whole environment, must conspire to make visual improvement conducive to survival and to ensure the consequent spread of the improvement across the generations.

Other things being equal, visual improvements are conducive to survival and reproduction. If the process was set up precisely to produce such improvements, to produce an organism capable of apprehending its environment, this is just what one would expect. For a superintending intelligence would be able to correlate all organic and environmental factors to ensure that, in the long run, light-sensitive organs would develop and reach a highly efficient form. But the Darwinian believes that there is no superintending intelligence. It just happens that mutations that construct organisms which reproduce more efficiently are conserved over time. The improbability of the mutational process lies in the fact that the mutations must be small and continuous enough, and correlated precisely enough with all the other factors, in body and environment, to produce stable, organised and reproducing molecular complexes. Theism and pure natural selection offer competing hypotheses to account for these facts. Theism says that they are just what one would expect, if there exists (in God) a consciously formulated goal of producing visual sensations in molecular complexes. Natural selection says that they are highly improbable, and are therefore highly unlikely to continue in any particular direction (for the continuation of the process is just as improbable as its existence in the first place). If it is rational to choose the hypothesis that gives a higher probability to a process, there is little doubt which hypothesis is to be preferred.

Dawkins works out in some detail a computer program which mirrors this computer simulation of the development of the eye. He calls it the 'Biomorph' program.[17] The program starts with a few basic instructions which generate outline shapes by extending lengths of lines, changing angles and so on. These instructions are followed 'at random'; that is to say, in no particular planned way. Nevertheless, the starting positions and the rules have been very carefully chosen to produce potentially interesting and complex shapes. It took Dawkins a lot of time and trial and error to get suitable rules in places. That is a very significant fact, when one is talking about 'randomness'.

As the program plays out, all sorts of shapes emerge. The computer operator then 'selects' from among all these shapes the ones in which she has an interest, ruthlessly discarding all the uninteresting or blotchy ones. She selects the shapes that seem to be beautiful; or which, in one case, look as if they might in time possibly make up the name 'Richard Dawkins'. After repeated selections, always choosing for survival the shape that comes nearer, by however little, to the pattern required, lo and behold, the pattern one wants arrives (or very nearly, in the case of Dawkins' name).

It is of the utmost importance that it is the mind of the human operator doing the selecting. And it is doing it with a specific goal in mind. This is not an unplanned or random process at all. The Biomorph program is as clear a demonstration as one could want that intelligence is required to get a desired complex state out of a computer program. What Dawkins suggests is that, in nature, the selecting is purely 'natural'. That is to say, it is the result of competition among the forms themselves, eliminating some and leaving survivors. But it is immediately clear that any such competition is very unlikely indeed to result in beautiful or meaningful forms. If the forms generated by the application of the rules (the laws of physics) are just going to be left to battle it out, so that the strong survive, then one will be liable to get a species of brutal, strong and ruthless predators, without any sentiment, conscience or sympathy for others. One will get the original predatory alien, the delight of sci-fi horror films. Or, even more likely, one will get nothing at all, since forms will keep mutually exterminating each other, and nothing interesting will ever develop.

The contrast between the Biomorph program and the course that would probably be taken by an unplanned struggle for survival between constantly mutating organic forms could hardly be greater. The former, being intelligently guided at every step, moves smoothly and progressively towards a desired goal. The latter is a recipe for chaos, anarchy and mutual extinction. What this shows is that, if natural selection were the only guiding principle of evolution, it would be unlikely in the extreme that any conscious beings would ever come to exist, much less that they would develop consciences and rational forms of thought. The Biomorph program shows precisely the implausibility of thinking that natural selection is the only factor in evolution, as compared with the hypothesis of a designing and selecting intelligence.

THE LIFE CODE: THE ARCHITECT'S PLAN

I have just pointed out that there is no survival efficacy in a mutation that only affects one cell or group of cells, however advantageously. In some way, mutations must correlate to produce an overall advantage for an organism as a whole. This fact, of the correlation of living cells to form an organism, draws attention to the fourth and fifth crucial stages in the evolution of life. So far, we have considered how simple sub-atomic elements build to form relatively stable molecular complexes (stage one). Some of these complexes form immensely complicated sequences of nucleic acids which begin to replicate themselves. These are the 'double helixes' of DNA, the structure of which has been unravelled only late in the twentieth century (stage two). Then, mutations occur in the components of DNA, resulting in a gradual 'improvement' in organic life-forms (stage three). DNA itself consists of long chains of four nucleotides. This in itself is of amazing complexity and organisation, but its real significance lies, not in its own structure as such, but in the fact that it is a 'code' or 'recipe' for building large, stable molecular complexes (bodies) and their appropriate forms of behaviour. It is this capacity of DNA to form a unique construction-code that is the fourth crucial stage in the evolution of life.

It is the unique characteristic of DNA 'recipes', which consist simply of long chains of four nucleotides, that they produce such things as living bodies with definite physical properties. The sequence of nucleotides in DNA forms a unique construction-code for embryonic development, for the construction of bodily forms and behaviour-patterns. Thus the eye exists in a body with a central nervous system, capable of reacting to information the eye provides. That is all part of the complex process that makes the development of the eye an improvement, even in terms of replicatory efficiency. DNA sequences are mixed and spliced in highly ordered ways, and mutations in the sequences produce variations in relevant physical properties. What one has is the assembling of a recipe for constructing physical properties in complicated interrelationships. These recipes are modified (mutated) by various external causal stimuli, not in a chaotic way, but in such a way that small changes in the relevant physical properties occur. As such modifications continue to occur in the process of replication, there is a cumulative development of such things as complex sensory organs.

When mutation is seen in this light, it is clear that it is not random, in the sense that it is quite unpredictable. It may well be that genetic change, like sub-atomic change, is probabilistic rather than deterministic. But the probability of a particular change can be specified accurately, as can the nature of the possible changes and the final outcome of a process of such changes. As Dawkins actually says, it is only random in that not every specific change is adaptive, or more pessimistically that few changes are adaptive. The overall process of mutation, however, is clearly adaptive, when taken, as it should be, in its total causal environment. The idea of randomness takes its emotive force from being considered in isolation from its context. But it cannot be properly so considered, since the nature of changes depends on that context anyway.

On this issue, then, there is no real difference as to the facts between Dawkins and a theist, even though Dawkins glosses over some of the deep disputes between biologists on matters of detail. The difference is about the interpretation of the facts. Dawkins, having described the amazing way in which DNA sequences provide coded information, which is translated into physical bodies and their behaviour, just as the plans of an architect provide information from which a building can be constructed, goes on to say, 'There is of course no architect.'[18] It just happens, by chance, that DNA sequences are recipes for building coherent stable bodies with sense-organs. It just happens that such things as cosmic rays cause mutations in those sequences which in turn cause just the right degree of mutation in the physical properties of bodily organs to engender a process of adaptive improvement. It just happens that the environment is favourable to the differential survival of such improvements, over thousands of generations. It is all truly amazing, and highly improbable, given all the ways the system could break down. The DNA messages could be irretrievably scrambled; they could fail to provide information for building bodies; mutations could be too large or erratic to allow natural selection to work; the environment could fail to support mutating life-forms at all (with the dinosaurs, it did fail). Nevertheless, it *could* all happen by chance, and to say that it did has the virtue of simplicity or parsimony, in that it eliminates all reference to a designing intelligence.

It is clear, on the other hand, that this is just the sort of system that super-intelligent computer programmers, or a God, might set up, if they wanted to produce highly adaptive conscious organisms from a fairly simple initial set of primary rules and array of fundamental particles. Simplicity is desirable, in scientific and in metaphysical explanations. But it is not the only intellectual virtue, and it turns into a vice if it is bought at the price of high improbability, or of omitting a significant range of relevant data. After all, the simplest possible hypothesis is that there is only one thing, or even, as Peter Atkins audaciously suggests, that there is nothing. But that is not a very convincing hypothesis to anyone who has noticed that, in actual fact, there are quite a number of things. This may lead one to think that simplicity should not be considered as just paucity of initial data.

The sort of simplicity a physicist is interested in is the sort possessed by Einstein's theory of general relativity, which contains no less than fourteen independent equations for describing behaviour in accelerating frames of reference. It is still a beautifully simple idea, however, in its symmetry and logical completeness.[19] The idea of simplicity here is that of integrated symmetry or harmony, relating diverse data under highly integrated principles. In this sense, God is a simple hypothesis, since the hypothesis provides one integrating principle for a vast range of complex and many-levelled data. One physicist suggests, for such a principle, the goal of creating 'high-level loving and sacrificial action by freely-acting self-conscious individuals'.[20] The evolutionary system is one the beauty and elegance of which suggests that it might well have a very wise architect. The hypothesis of God both makes the actual system highly probable, if God desires such a system to exist, and guarantees its continuance into the future (or at least guarantees that its purpose will not fail).

NOTES

[1] Richard Dawkins, *River out of Eden*, pp. 151–61.
[2] Dawkins, *The Selfish Gene*, p. 14.
[3] Eugene Wigner, 'The Unreasonable Effectiveness of Mathematics'.
[4] I have developed this idea of God in greater detail in *Religion and Creation*.
[5] Dawkins, *The Selfish Gene*, p. 16.
[6] Dawkins, *The Blind Watchmaker*, p. 139.
[7] Ibid., p. 307.
[8] Ibid., p. 41.
[9] Ibid., p. 306.
[10] Ibid., p. 312.
[11] Ibid.
[12] Ibid.
[13] Ibid.
[14] Dawkins, *River out of Eden*, p. 78.
[15] Ibid., pp. 78–83.
[16] Ibid., p. 80.
[17] Dawkins, *The Blind Watchmaker*, ch. 3 and Appendix.
[18] Dawkins, *The Selfish Gene*, p. 24.
[19] S. Weinberg, *Dreams of a Final Theory*, p. 107.
[20] G. Ellis, 'The Theology of the Anthropic Principle'.

CHAPTER SEVEN

Evolution and Purpose

THE UNIVERSAL WEED: WHY IT DOES NOT EXIST

But why would God desire such an evolutionary system? Would it not have been easier, and possible, for God to have created whatever God wanted instantaneously, as theists have often believed in the past? At least, should God not have made every mutation a favourable one? I have already suggested that the reason God creates is to actualise a distinctive sort of goodness, or set of intrinsically worthwhile states. There is one distinctive sort of goodness which lies in a process of creative learning, self-shaping and self-expression by conscious agents in community. A community of conscious beings, emerging from and gradually learning to understand and master a material environment, can actualise a distinctive set of values. If such beings are to possess creative freedom, the basic processes of nature must be non-deterministic, to allow the openness within which freedom can operate. That is, it must not be the case that every state of the physical universe is sufficiently caused by some previous state plus a set of physical laws. If X is the sufficient cause of Y, then if X exists, Y must exist, and no other state than Y can exist. In a determinist account, every state of the universe (except the first) is necessarily what it is, and comes into being by the application of some finite set of laws to some prior state. When I say that the processes of nature are non-deterministic, I mean just to deny this determinist account.

I am not thinking primarily of quantum indeterminacy, which normally operates in a very limited sphere. Some scientists indeed hold that quantum indeterminacy, plus the multiplier effects of some physical systems, does give rise to a significant indeterminacy in nature. Paul Davies, for instance, writes: 'The intrinsically statistical character of atomic events and the instability of many physical systems to minute fluctuations, ensures that the future remains open and undetermined by the present.'[1] But I do not wish to confine the sort of creative openness I have in mind to such events. Nor am I denying that many physical laws are deterministic, in the sense that, for instance, the solution of the differential equations used in physics is wholly determined by the values of the relevant variables. What I am denying is that one can ever state a set of laws such that, taken together with a description of some initial state of the universe, they will entail a complete description of every subsequent physical state of that universe.

This is, of course, a hypothesis that could in principle be disconfirmed. The reason for postulating it is that it is a condition of the reality of creative freedom, which seems to exist, at least at the level of human consciousness and activity. The natural sciences are nowhere near being able to disconfirm the hypothesis, and there is good reason to think that they will never be able to do so. For the precise specification of initial conditions must eventually run up against the Heisenberg uncertainty principle, and this places limits on the amount of information one can obtain. Since determinism can never be proved, determinists are in an exactly similar position to theists. They have to postulate a hypothesis, largely on faith. The main reason for postulating the determinist hypothesis is probably the claim that it is a condition of complete scientific understanding, when everything could be explained by reference, preferably, to a few simple laws and a simple initial state. Further, it may be said, deterministic laws, like Newton's laws, do apply within limited areas, which may seem to suggest that they apply universally.

The element of indeterminism involved in the 'freedom

hypothesis' is simply that not everything that happens is the result solely of the operation of a general law, or combination of general laws, upon some previous physical state. Such indeterminism, or at least the appearance of it, is commonplace in ordinary human affairs. If the majority of voters in the United States elects John Smith for President, there is no known general law that entails that result. The hard-line determinist may say that, if one gets down to the details of the brain-states of all the voters, one may find that those states are all determined by previous brain-states, and, when added together, they in fact entail the election result. But that is precisely the sort of dogma that is being questioned. It is at least equally plausible to think that voters decide for a great variety of reasons, and that many decisions are not entailed by prior brain-states. One can frame generalisations – that voters are likely to vote in their own interests, that only so many per cent of them will vote at all, or that the best-looking man will be elected. Some generalisations will be highly probable and others less so, but there will be many details that remain unconsidered, and such details will sometimes be critically decisive.

In human affairs, general laws allow for alternatives. They are not sufficiently determining. Even where laws themselves give determined results, it is by no means obvious that the laws cover every actual aspect of the phenomena in question. Newton's laws, for example, are abstractions which isolate measurable properties like mass, position and velocity, and predict how they will interact, if not influenced by any other factors. But in the real world there may be other properties than these, not all of them measurable precisely, which influence the future in various ways.

The indeterminist claim is not, therefore, just that various funny things happen at a sub-atomic level. It is that the formulable laws of physics – the algorithmic compressions with the aid of which we describe the flow of natural processes – are in fact incapable of describing fully the fuzzy, complex and unique aspects of the natural world, which influence to varying degrees how that world will develop. The physical world is not bound by chains of measurable and universal regularity. It is quite remarkable that algorithmic compressions are possible, and they illuminate our understanding of the universe enormously. But to think that such shorthand algorithmic devices give a complete and exhaustive description of how the universe changes, in all its details, is to abandon all thought of emergence, uniqueness and creativity, the most evident features of the very human consciousness that has generated the algorithmic compressions, the 'laws of nature', in the first place.

If this is so, even before the level of conscious freedom is reached, non-deterministic processes must govern the emergence of life and consciousness. If such processes are non-determined, even by God, and if that is their essential nature, then God plainly cannot ensure that they all issue in what God would regard as favourable outcomes. God could, however, devise natural processes of change that would be open to a continuing divine influence, and that would be statistically oriented towards developing self-organising complexity.[2] That strongly suggests a probabilistic account of genetic mutation – what a biologist, looking simply at the physical processes themselves, might well call 'randomness'. Again, theism is seen to have great theoretical elegance in explaining why the mutational process should be 'random', and yet how such an apparently random process should move in what seems to be the constant direction of self-organising complexity. For a process directed towards the emergence of free sentient beings must inevitably move in the direction of greater complexity, while its non-determinism allows many particular probabilistic deviations from that direction.

It follows from all this that natural selection cannot be the sole explanation of evolutionary change. Natural selection, the cumulative development of more efficient replicators in a competitive situation, is certainly an important factor in evolution. If evolution is to occur at all, organisms must replicate efficiently and they must mutate in the direction of greater complexity and integration of structure. It is by no means obvious, however, that the more efficient replicators will be those with such greater integration of structure. In other words, natural selection is a necessary but not a sufficient condition for emergent evolution. In fact, it would be entirely reasonable to think that, on the principle of natural selection alone, very complex and delicately integrated structures would be less likely to replicate efficiently, since they might break down more often, and be more open to attack by simpler but more virulent organisms. The most efficient replicator might be something like a giant poisonous weed, or at least a number of such weeds in symbiotic competition, which kill off all incipiently more complex forms, strangling them before they can take root.

In fact, of course, things have not happened like that. What has happened is that one-celled entities have acquired the capacity to split and develop, so that they do not simply replicate. Rather, their duplicates develop during embryogenesis in entirely different ways to form highly differentiated bodies. How each cell develops depends upon its bodily environment. So millions of cells, with identical DNA, develop within maturing bodies in quite different ways, and in perfectly balanced interaction with each other, to form coherent organisms. This is the fifth crucial stage in the evolution of life. Not only do the intertwined chains of DNA form a code for building bodies, but each cell in those bodies develops in a way dependent upon its place within the whole organism. It looks as though the nature of the whole determines the nature and function of its parts. The fifth stage is a stage of holistic codetermination. According to this holistic principle, the development of the individual cells of an organism is closely correlated with, and looks as though it is determined by, the needs of the whole organism. Professor Goodwin thus argues that 'organisms are as real, as fundamental, as irreducible, as the molecules out of which they are made. They are a separate and distinct level of emergent biological order.'[3] In a tiger, for example, some cells develop into sharp teeth, some into intestines suitable for digesting meat, some into strong legs, and so on. The teeth would be useless without intestines able to cope with meat. The carnivorous digestive system would be useless without the means of catching prey. All these parts interrelate, and it is implausible in the extreme to suppose that their interrelation is fortuitous. There is a co-ordination between the way different sequences of DNA are 'turned on' to build different bodily parts, each one of which only works in conjunction with all the others. At this point it does look as though quite a new principle is at work, the holistic principle that the way a part (a cell) develops depends upon the whole of which it is a part. The complex organism constrains the behaviour of its parts. This is the first clear example in the natural world of what some biologists call 'top-down causation'.[4] The DNA is coded, not only to translate into sets of physical properties, but to translate into the development of a harmonious and integrated unity of cells which together form one organism, and which develop their potential only within such an organism.

THE GREAT REVERSAL

These facts provide a very strong challenge to Dawkins' theory of the selfish gene, the theory that, insofar as one can speak of any point in evolution, that point is simply the survival of bits of DNA (genes being small sequences of DNA). Dawkins supposes that, if one looks at living organisms, one can ask what their Utility Function seems to be. That is, if they had been wisely engineered, or designed for some purpose, what would we infer that purpose to be, simply looking at their present nature?

'We now understand the single Utility Function of life in great detail,' he writes.[5] It is the survival of DNA. 'Everything makes sense once you assume that DNA survival is what is being maximised.'[6] This is what one might call the Great Reversal in Dawkins' philosophy. In fact, the survival of genes is not maximised by evolution, since the whole process proceeds precisely by the mutation of genetic material, that is, by replacing genes by better ones. Moreover, every specific gene has a very short life, so that what endures is the coded information a DNA sequence carries, not the material stuff itself. Even on Dawkins' own principles, the Utility Function is not the survival of the bits of matter that make up DNA sequences. Nor is it the unchanged survival of the coded sequences themselves. It is the gradual rearrangement of DNA codes, to form ever-more sophisticated programs for building bodies. But it looks distinctly odd to say that what an engineer might aim for is the emergence of programs for building bodies, while the actual existence of the bodies themselves is an irrelevant by-product. This is the Great Reversal – the only really intelligible goal is seen as an unintended and unforeseen by-product, while the coded information containing the program for achieving the goal is seen as the goal itself. It is just like saying that the important goal of cookery is the production of recipes. The cakes themselves are unintended by-products of the recipes. Something has gone seriously wrong!

The Utility Function of life is not the maximisation of unchanged replication of DNA sequences at all. It is the gradual reorganisation of the genetic program, the software coded into DNA sequences. But what is the software for? It is for building more and more complex bodies. In itself, the nucleotide sequences are meaningless. But as recipes for body-building, they have a distinct Utility Function. The Utility Function of life, on Dawkins' own principles, is the building of more complex integrated bodies. We might not yet see the point of that – we shall see in a later section that the reason for having complex bodies is to allow the development of consciousness – but it does seem to be what Dawkins' own imaginative experiment of 'reverse engineering' would throw up, as what natural selection seems to be maximising.

Bodies, however, for Dawkins, are of little importance. The important things are the genes, which are selfish, in seeking their own survival by any means. Thus 'we are survival machines – robot vehicles blindly programmed to preserve the selfish molecules known as genes'.[7] The genes 'swarm in huge colonies, safe inside gigantic lumbering robots, sealed off from the outside world, communicating with it by tortuous indirect routes, manipulating it by remote control'.[8] By an incredible inversion, the body becomes a blind robot, while the genes take on personal qualities, communicating with and manipulating the world, to realise their wholly selfish purposes.

Of course, Dawkins knows that genes are not little conscious manipulators. All this is said for emotive effect, to downgrade the importance of personal consciousness and agency. But it is worth noting how the arch-reductionist constantly speaks in terms of purpose and agency, even while explicitly denying its existence. 'We humans have purpose on the brain', he writes; it is 'a nearly universal delusion'.[9] If it is such a delusion, why does he constantly speak in terms of Utility Functions, selfish genes and evolution as a 'scramble for selfish gain'?[10] The plain fact is that consciousness and purpose do exist, at least in the higher animals. It is therefore not absurd to ask whether purpose is not rooted in the physical structure of being itself. It is not absurd to suppose that there may be a purpose in the evolutionary process. But it is absurd to think that it can be found in small unconscious bits of DNA, pursuing their 'selfish' goals. Any purpose there is will not be found within the physical structure of unconscious entities. Purpose must be founded in some consciousness, and the obvious place to locate it is in the mind of the cosmic architect. If one locates a universal purpose in a cosmic mind, it is highly unlikely that the purpose would be simply to maximise DNA replication for its own sake.

THE ALTRUISTIC GENE

Genes would in fact, on a theistic hypothesis, be inherently unselfish. That is, bits of DNA would exist, not for their own sakes or simply to replicate themselves, but in order to build bodies which could at some stage contain consciousness, capable of apprehending and creating intrinsic values. Genes would still need to replicate successfully, even though they would be constantly improving by mutation, but that does not make them particularly selfish. In fact, it is Dawkins himself who shows very clearly how unselfish genes actually are.

'The manufacture of a body is a co-operative venture,' he writes.[11] It cannot be the case that every gene competes with every other, seeking to eliminate it, for in fact 'selection has favoured genes which co-operate with others'.[12] At this point, the committed Darwinian constructs a theoretical epicycle to explain away the appearance of altruism, of co-operation between genes, in building bodies. The epicycle is that such co-operation will itself maximise gene replication. If what will be replicated is a specific sequence which forms a recipe for building some part of a body, not a material entity, it does not matter if that sequence is in a different physical cell or not. So a gene may best replicate a sequence it contains by co-operating with other genes containing the same sequence, or by co-operatively building defensive machines (bodies) in which they can all coexist. Now the metaphor of 'selfishness' has been pushed beyond usefulness, for each gene strives to enable any sequence like the one it contains to replicate.[13] That is, at the very least, a limited altruism. In fact co-operation becomes a much more important characteristic of genes at the phenotypal level (the level of body-building) than any interest in self-survival.

But one must push this epicycle a little further. What genes do is not simply replicate their nucleotide sequences exactly. They mutate, and thereby produce at least some better replicating sequences. So now what genes are doing is co-operating to produce improved sequences – that is, sequences containing codes for building better surviving bodies. Each gene gives up its individual life by co-operating with many others to produce superior versions. This co-operation extends to the environment as a whole, too, which co-operates in the sense that it favours well-balanced and positively interacting organisms. At that level it is called 'adaptation'. If one is going to speak in personal metaphors about genes, it is better to speak of co-operative self-sacrifice than to say that, 'at the gene level, altruism must be bad and selfishness good'.[14] But perhaps it is better to drop the metaphors, and just say that sequences of nucleotides in DNA are coded so that they build highly integrated bodies. If there is a Utility Function in evolution, a design or purpose, it seems to lie in the construction of such bodily forms.

But why should bodies exist? Dawkins' Great Reversal asserts that bodies are machines for the survival of genes. 'We animals are the most complicated and perfectly designed pieces of machinery in the known universe,' he writes.[15] But the preservation of bits of DNA 'is the ultimate rationale for our existence'.[16] As a matter of fact, nothing more pointless, no less convincing rationale, could be imagined for the existence of bodies than this. Who could give a fig for the survival of bits of DNA? Not genes, which know and care for nothing. Not us, who might be interested in personal survival, but would be perfectly happy if that could be accomplished without any DNA at all. Not God, who can hardly find intrinsic worthwhileness in the existence of strings of nucleic acids. To put it bluntly, nobody cares about strings of nucleic acids at all!

There are two qualifications to put on this remark. One is that God may appreciate the beauty and elegance of the structure of DNA, and in that sense it has intrinsic worth. Somebody – God – values it for its own sake and is glad that it exists. It is important to see that DNA only has such worth when it is actually appreciated. Without any conscious recognition, it would have no worth at all. Its intrinsic value is its value-when-apprehended-by-some-consciousness, if only that of God. The other qualification is that, of course, *we* care about DNA, if it is a necessary condition of our existence. But we then care about it, not as an intrinsic good – one that is good for its own sake – but as an instrumental good – one that is necessary for some intrinsic good to exist.

THE IDEA OF INTRINSIC VALUE

This helps us to see that, if there is any point or purpose to the universe, it must eventually lie in the existence of some thing, state or process that is intrinsically good. The question about the purpose – the Utility Function – of life, is thus the question about the intrinsic goods that life realises or makes possible. Such goods entail the existence of consciousness, so they must relate to conscious experience in some way. For nothing can be called intrinsically valuable unless it is actually valued by some conscious being. One of the conscious states that beings put value on, just for their own sake, is the contemplation of beauty, whether in music, painting, poetry or nature. Looking at the Alps in the snow and taking pleasure in their colour, grandeur and austere magnificence is something that we can enjoy just for its own sake. Such moments of contemplation are, for many, among the high points of human experience.

At just this point, however, Dawkins puts his foot down. 'Beauty is not an absolute virtue in itself,' he writes.[17] He selects the one thing that seems to me to be an intrinsic value if anything is – the appreciation of and the happiness that comes from contemplating something of beauty. And he simply denies that it is an absolute virtue. In face of that, I must admit I am speechless. But I do not think he really means it, since he places so much emphasis on simplicity and beauty in scientific theories that he must regard it as a virtue. However, one may see here another example of the sociobiologist's upside-down view of the world, which says that all obvious values are not values at all, but by-products of physical processes. This is a *non sequitur*, since something can be both a by-product and a great value. But it raises a more serious issue.

One of the most destructive ploys of atheism is to suggest that facts are there, in the outside world, while values are just subjective reactions, which vary from one person to another, matters of personal taste. If you think torturing babies is good, that is a matter of taste. It is not objectively wrong to torture babies. The crassness of this view is obvious. What is needed to counter it is the simple reflection that there are some things that every rational person desires and values. These include food, shelter, clothing, the basic necessities of life. They also include such things as a degree of happiness, of friendship, of knowledge and of freedom, since without those things one could not enjoy anything with any sort of security. Among these desirable states, some are instrumental, they are means to other values (as health is a means to doing things we want). Some are intrinsic, desirable for their own sake (pleasure is the most obvious of these). Of course there are many disputes and dilemmas in morality. But that should not prevent one seeing that there is a fairly clear and basic set of intrinsic values, of states any rational person would desire for their own sake. When we speak about 'human rights', we are not speaking about some purely personal preferences which vary from one country to another. We are speaking about giving everyone access to some share in the intrinsic values that make human life worthwhile.

Torturing babies is wrong because it not only deprives a human life of intrinsic values, it actually causes states of pain which any rational person would fear and seek to avoid. Ethics is the immensely complicated task of working out exactly how to obtain the greatest range of values, as fairly distributed as possible, among as many conscious beings as possible. But the basis of ethics is simple. It is the specification of those basic intrinsic values that all rational beings would desire. It is in this sense that intrinsic values are objective. They are states that all sentient beings have a good reason to want. Most ethical dilemmas arise because of the difficulty of deciding how the possibility of getting such states should be divided amongst various groups of people. In particular, it is often hard to decide how I should balance getting such good states for myself and providing them for other people. There are plenty of moral dilemmas. What is not in question, however, is that there are some objective values which provide obviously good reasons for action. If any being has a rationally defensible purpose, it must lie in the production of some value, and ultimately in the realisation of some intrinsic value, which can be enjoyed by some conscious being.

Now what is a value for Dawkins? Maybe he thinks there is no such thing, that what one person likes another does not. There are only subjective likes and dislikes, and one should not speak of value at all. Perhaps he thinks that, because there are different forms of beauty (different sorts of music and painting, for example), there is no such thing as beauty at all. Perhaps he thinks that the appreciation of such forms is not something everyone has a good reason to desire and aim for. He would, I think, be doubly mistaken. Though there are different sorts of music, each sort has definite standards of excellence and appreciation. The good reason for trying to understand at least some of these is that it will bring a high degree of satisfaction and pleasure, and will in the end prove to be self-justifying.

In any case, a value is always a reasonable end or purpose of action. If you ask, 'What was your purpose, or end, in listening to that music?' a perfectly satisfactory answer is to say, 'My end was to appreciate its beauty, for its intrinsic value.' To give a rational end or purpose is to give some intrinsic value. But Dawkins does not speak in this way. What he says is that 'existence and survival were only means to an end. That end was reproduction.'[18] He selects as an end something of no value at all, something that no rational being could choose just for its own sake. That is because he interprets an 'end' as just the property that is maximised by a process, without any valuation of it at all. Existence and personal survival are not maximised by evolution, since all individuals die. But reproduction is maximised, since the most efficient reproducers do reproduce well.

A rational goal is something a mind could consciously aim at. A mind could hardly aim at reproduction for its own sake (without even considering the pleasure it may involve for animals). If reproduction really is the property maximised by evolution, that may seem to suggest that evolution is not designed by a rational mind. But if one reflects for a moment, it becomes clear that a goal of intrinsic value is not the same thing as a maximised property. Consider a symphony. From one point of view, what is maximised by a symphony is the number of notes played. But anyone who suggested that the goal a composer aims at, in writing a symphony, is to include the maximum number of notes would be totally wrong-headed. What is aimed at is the total sound-pattern, and there is only one such totality. So the maximised property is not identical with the rational aim, though one would expect that the property (the number of notes) is in some way necessary to accomplishing the aim. In the same way, reproduction (not necessarily the maximum possible) is necessary to the generation of minds that can appreciate beauty. Even if there is only one such mind, *that* will be the rational goal, not the number of reproductions necessary to achieve it.

Now we are in a position to see what the purpose of the DNA-coded construction of bodies is for, what its rational goal is. It is certainly not reproduction, or the survival of pieces of genetic code, which are not, as such, rationally choosable goals at all. What complex bodies do is make possible the development of central nervous systems and then of brains, which can receive information from the environment and respond to that information to achieve desired states. Bodies are not primarily machines for carrying genes. They are not machines at all. They are the generators and carriers of central nervous systems, networks of conscious interaction with the environment, making possible understanding, contemplation, happiness and rational agency. The purpose of genes is to build bodies, the purpose of bodies is to build brains, and the purpose of brains is to generate consciousness and purpose, and with them, for the first time in the history of the cosmos, the existence of intrinsic values. If there is a rationally choosable purpose, a Utility Function, in evolution, this is it. We have arrived at the sixth crucial stage in the evolution of life.

NOTES

[1] Paul Davies, *The Mind of God*, p. 201.
[2] The idea of a statistically directed and non-deterministic process is helpfully outlined by the mathematical statistician David Bartholomew

in *God of Chance.*

[3] Brian Goodwin, *How the Leopard Changed its Spots*, p. xii.

[4] See Arthur Peacocke, *Theology for a Scientific Age*, pp. 157–60. Peacocke is not himself referring to the development of organisms, but it would be odd if their behaviour was constrained by top-down causation, while their coming into existence as complex organisms was not.

[5] Richard Dawkins, *River out of Eden*, p. 105.

[6] Ibid., p. 106.

[7] Dawkins, *The Selfish Gene*, p. x.

[8] Ibid., p. 21.

[9] Dawkins, *River out of Eden*, p. 96.

[10] Ibid., p. 121.

[11] Dawkins, *The Selfish Gene*, p. 25.

[12] Ibid., p. 50.

[13] The metaphor of the 'selfish gene' has been castigated by Mary Midgley in two articles: 'Gene-juggling' and 'Selfish Genes and Social Darwinism'.

[14] Dawkins, *The Selfish Gene*, p. 38.

[15] Ibid., p. xi.

[16] Ibid., p. 21.

[17] Dawkins, *River out of Eden*, p. 120.

[18] Dawkins, *The Blind Watchmaker*, p. 200.

CHAPTER EIGHT

Brains and Consciousness

THE MYSTERY OF CONSCIOUSNESS

Consciousness is, as Dawkins says, 'the most profound mystery facing modern biology'.[1] It is more than that, however. It is a mystery that biology can never solve, because it is not a biological mystery. The mystery is how it comes about that the construction of brains, of complicated collections of purely physical particles, gives rise to something apparently non-physical: thoughts, feelings, dreams, images and intentions. Up to this point in evolution, it is just possible for the Darwinian to hold that there is nothing more to the cosmos than the interactions of physical forces. However improbable it may be, it is theoretically possible that brains might just develop by chance out of the fact that molecules tend to stick together in complicated lumps. But no collection of physical lumps can add up to even one simple and momentary feeling of pleasure.

Conscious experiences are radically new elements of reality, which seem only to come into existence when a certain stage of physically complex structure exists. We do not know why the firing of neurones in the brain should give rise to particular sensations or thoughts, though we assume that they do. Just as nuclear particles came into existence in the first seconds of the universe, so thoughts come into existence after about fifteen thousand million years of cooling and expansion in the universe. With thoughts, values and purposes come into existence. Life-forms with brains can cause changes in their environment that will give them things they value or desire. This is the most radical change in the ontology of the universe. It cannot be explained simply by appeal to physical laws and the way they operate on material elements. Such laws and elements, far from explaining conscious states, do not even mention thoughts and feelings. Here a new level of explanation is required, and it cannot be one solely in terms of physical factors. Conscious states cannot be explained in terms of physical properties alone.[2] Yet conscious states depend for their existence upon complex physical structures. The obvious conclusion is that the complex structures must be explained in terms of the conscious states they seem developed to produce. The existence of consciousness is the refutation of materialism. It presents one with facts no materialist account could ever explain, in principle. Yet the whole history of evolution seems superbly well designed to lead to the existence of consciousness. It is designed, in other words, to lead to levels of explanation and reality beyond itself.

At this point, the hypothesis of theism is not just superior to that of materialism; it is the only hypothesis with a hope of explaining the facts. What could be meant by explaining consciousness? Explanations in physics consist in providing simple laws in accordance with which complicated realities can be seen to develop out of simpler ones. With regard to consciousness, if we apply the same general form of explanation, what is required is the provision of a few simple principles in accordance with which very complicated conscious phenomena can be seen to develop, in particular physical environments. These principles will basically be those of value and disvalue – more simply, pleasure and pain. The fundamental form of explanations of consciousness will be purposive explanation – an explanation of how particular feelings and moods relate to the gaining of valued states and the avoidance of disvalued states. A purposive explanation explains the behaviour of an organism in terms of the goals it seeks or the states it seeks to avoid, that is, in terms of its purposes.[3]

If one is seeking an explanation for the existence of consciousness itself, that explanation can only be given in terms of the purposes it realises, or the intrinsic values it can apprehend. Theistic explanation seeks to show the intrinsic values that evolved organisms can realise and apprehend, and to show that this is an intelligible purpose of a conscious, purposing reality. The existence of minds can only be explained, if it can be explained at all, in terms of the intrinsic values that they can create and apprehend. The ultimate explanation of a plurality of dependent minds, which have originated through a long process of development in a material universe, lies in the existence of a mind that does not depend on such processes of development or on anything else whatsoever. It will be itself of supreme value, and thus self-justifying or self-explanatory, and it will generate finite minds for the sake of the goodness they can realise. This is precisely the hypothesis of God, which can then explain why physical elements arrange themselves in such improbable and complex ways – in order to generate conscious states capable of understanding and modifying their physical natures.

WHAT DOGS SMELL

The evolutionary naturalist sees consciousness, feeling, intention and evaluation as by-products of a mechanistic process, without any causal effectiveness and without any positive role in a fundamentally physicalist universe, a universe whose basic reality is that of physical particles interacting in accordance with a few general laws. It is pretty hard to see how such mechanical interactions can give rise to consciousness. If the basic physical particles only have such properties as extension, mass, velocity and position, how can the quite different properties of colour, smell and touch arise from them?

Neuro-physiology can show how smells, for example,

result from the inhalation of molecules, which impact on nerve-receptors within the nose, causing electrical currents which travel down the nerves and stimulate a specific part of the cortex. Few would be tempted to say that the smell *is* just the molecules, since it is obvious that the smell only occurs when the cortex is stimulated, and by that time the molecules have no direct contact with the brain. There is some temptation to say that the smell is just the excitation of a specific part of the cortex (what is called central-state materialism). This view is strengthened by the fact that direct stimulation of the brain can produce sensory impressions in patients undergoing brain surgery. In principle, therefore, a brain in a vat could be stimulated artificially, and could have all the sensory impressions of a normal human being, even though it had no body or senses at all.

It is an experimentally established fact that the stimulation of specific parts of the brain results in the occurrence of specific sensory impressions – and the stimulation of other parts of the brain may result in thoughts, images or feelings. But it is an almost total mystery how this can be. We can observe a human brain and see it being stimulated. We can record all the physical data – the electrical current, the interaction of synapses, the movements of electrons and so on. All the physical forces can be observed and described, measured and recorded. But what we cannot observe is the smell that these forces bring about. To say that the smell *is* these forces seems absurd, since we can record the forces exhaustively and know absolutely nothing about the smell.

How do we know a specific brain-state results in someone smelling something? Only because the person tells us so. We have to ask. Otherwise we have no idea whether the person is smelling, touching, seeing or dreaming, or is completely unconscious. Of course, once the person has told us that stimulation X results in smell Y, we can infer that, whenever stimulation X occurs, smell Y occurs. But that is still an inference. We can never confirm it from our own experience.

So, if we stimulate the brains of animals, we cannot know what sorts of sensation they are having. Dogs have many millions of brain-cells devoted to the sense of smell, and we know from their behaviour that they can detect many more smells than humans can. But we have no idea of what it is like for a dog to smell. All we can do is construct models of what it might be like for a dog to have a very rich array of smells. We are working from analogy from our own case, assuming that similarity of brain structure and of behavioural response implies some similarity of experience. Such analogies are quite compelling. After all, we cannot get inside another person's brain to see what they are experiencing, but we assume it is roughly like what we experience. In the human case, language provides an additional clue, as well as behaviour and brain structure. But the fact remains that we are limited to inferring what the experiences of other brains are like from our own case. The world of consciousness seems to be different in kind from the physical world, though in some mysterious way it is interfused with it. We all have access to the physical world, but none of us will ever know just what a dog smells. We just have to guess.

WHY THE COMMONSENSE WORLD SHOULD NOT DISAPPEAR

Some philosophers have complained that such an inference is just too big. We only have one case to go on – the correlation of our own brain-states and our own experiences. That cannot possibly justify us in assuming so many thousands of allegedly similar correlations, none of which we can ever check for ourselves. As a matter of fact, of course, we virtually never know about any correlation between our own brain-state and an experience. We hardly ever know anything at all about our own brain-states. So that is not the sort of inference that takes place in real life. We simply operate in a world that we assume gives us reliable knowledge of an objective environment and of the thoughts and feelings of other persons. The commonsense world is a world of experiences. It is the scientific world of molecules and brain-states that is the result of very sophisticated observation and theory, and we believe in it because we trust the reliability of science.

This simple and obvious point is so important that it needs to be stressed as strongly as possible. The place we all start from, in thinking about the nature of reality, is a world of colours, shapes, smells and feels, of thoughts and feelings, of experiences which we naturally and unthinkingly interpret as conveying meanings and purposes that come to us from outside ourselves. We are not passive spectators of some sort of internal TV screen – that picture itself derives from later scientific theory. We are agents, from the first responding to the thoughts and purposes of others – our mothers first of all, then our families and friends – in an objective, interacting world which is a rich arena of thoughts, feelings and sensations intertwined one with another.

It is because of this that theism, of some sort, is an entirely natural and unforced belief for human beings. Since we do not separate thoughts, feelings and sensations from one another in our basic experience, it is a natural proclivity to interpret the objective world as conveying thought and purpose, as communicating with us for good and ill, as expressing the character and purpose of some conscious and intentional being or beings. It is as natural to interpret the world as communicating the will of God, or perhaps of many gods, as it is to interpret parts of our sensory impressions as communicating the purposes of other people. This is not a theoretical justification of God. It is not theoretical or reflective at all. But it is a reminder that belief in God is not a strange and difficult hypothesis, inferred by some sophisticated argument from our experience of the world. Belief in God is an immediate and natural interpretation of experience as communicating an underlying personal reality, which is like us in some fundamental respects.

Theism is a very natural part of direct realism, the assumption (not the reflective hypothesis) that we immediately experience things as they are, so that reality is colourful, tactile, beautiful, meaningful and purposive, just as we immediately take it to be. For direct realism, beauty, meaning and purpose cannot be abstracted from some alleged purely sensory experience, and attributed to the perceiving mind. It is the idea of a perceiving mind that is an abstraction. The direct realist sees beauty and meaning in the things themselves, to be profoundly or superficially, deeply or shallowly apprehended. God, at this level, is the widest and most basic reality which expresses beauty, meaning and purpose – and also terrible power, mystery and apparent arbitrariness – in all the particular aspects of our experience.

Of course, direct realism cannot survive the onset of reflection, which begins the process of abstraction, generalisation and theorising, to provide a deeper understanding of our lived and experienced world. Reflection can clarify the underlying structures and intelligibility of the world of human experience. It can explicate and systematise the idea of God which is implicit and uncritically present in phenomenal experience. But one thing reflection cannot do is undermine the richness and reality of that experienced world itself. A first principle of a truly explanatory theory is that it must retain the full reality of the phenomena it is seeking to explain, and not eliminate it or transform it into something else. It is exactly that principle that is contravened by materialism. Materialist theories end up by denying the very realities of conscious experience they set out to explain, and in that sense they are not true explanations at all.

THE THREE WORLDS OF HUMAN EXPERIENCE

As one begins to reflect on experience, one wants to know why we smell things as we do. The direct realist supposes that, as we move around the world, we detect smells by sniffing them. The smells are already there, and we pick them up with our noses. When we begin to analyse experience, we discover a causal story connecting molecules to brain-states, and we proceed to construct a very different picture of the objective world from that of direct realism. Now the 'real world' consists of molecules, interacting with one another and with atoms and electrons in the brain. Only when the very complicated structure of the brain exists will smells come into existence. Those smells are projected onto the world in some way, but they are not actually there, until some brain-state exists. If we ask where the smells are, it now seems that they are not in the physical world at all. We have to invent another world, the 'mental world', which is a sort of private model of the physical world, with very different properties. Nobody else can get into our private mental world of experience, but it models the physical world in a rather exact way, yet without revealing to us the 'true, atomic, structure' of that world at all.

It is important to see that the physical world is a mental, theoretical construct. It is not the world we seem to experience and act in. It is a sort of hidden, yet objective, reality which in some mysterious way gives rise to the experienced reality that is first known to us, and provides the basis of our theories. It has the peculiarity that it forces our experienced world to become a purely private world, which no one else can enter, and it forces the objective world to become one that no one ever experiences as it truly is.

This is very unsatisfactory, from a theoretical point of view, since it is the experienced world that seems to us to be objective, and what is now said to be the objective world seems clearly to be a theoretical construct. The materialist takes a short way with this theoretical difficulty, and simply eliminates the experienced world, or at best describes it as a causal by-product of the physical world. In this way, theoretical simplicity is obtained at the cost of denying the very experiential basis of the theory. The purely theoretical becomes the truly real, and that which is experienced as real becomes merely an illusion. Materialism represents the triumph of theory over experience. But the triumph is a hollow one, since it is only experience that allows the theory to exist in the first place.

A more satisfactory hypothesis might be that there are not two worlds, a 'real' public, colourless, purely law-governed world and a private, subjective or illusory and value-laden one. The phenomenal world is how things appear to sentient beings from a certain point of view within it, the point of view of interacting agents who seek value and meaning in their shared communities. The physical world is how things appear to the same sentient beings from a different viewpoint, a dispassionate and analytical viewpoint which attempts to exclude all elements of value and purpose and to expose the general regularities and fundamental powers that set limits on the interactions of things. There is just one world, seen from viewpoints that express different interests in it and modes of relationship to it. The phenomenal viewpoint is not 'less real' than the physical viewpoint. Nor is it some sort of surprising by-product of a purely physical world.

To say that the physical world is a theoretical construct is not to say that it does not really exist. It seems clear that, long before phenomenal consciousness of any sort we can recognise existed, there was an objective reality. Indeed, it was just such a reality that, through emergent evolution, brought about the complex physical conditions necessary for phenomenal awareness. The point is, rather, that the way we conceive of such a world is derived by abstraction from our phenomenal experience, and so cannot be used to undermine or deny that experience. As Bishop Berkeley pointed out centuries ago, what we call the physical world is constructed out of a subset of perceived properties – the so-called 'primary qualities' of shape, extension, mass and position – which have been abstracted from the total array of properties humans apprehend, largely because they are measurable and verifiable by virtually all observers.[4] This physical world is no less real for being a mental abstraction. It has often seemed to have a sort of enduring solidity that made it a bedrock of reality, something that one could call 'truly real'. In the light of its reassuringly solid reality, the momentary and transient data of the senses were often relegated to a realm of half-reality. Such a relegation is quite unjustified. The mind has taken its revenge, and the physical world of primary properties has itself been shown to be less solid and basic than was once, in safe Newtonian times, thought. In recent years our conception of the physical world has undergone a fairly radical revolution, because of the advent of relativity and of quantum mechanics.

When the physical world could be said to consist of indivisible atoms, interacting in accordance with laws of mechanics, at least it had a feel, however illusory, of solidity and stability. In the quantum world of modern physics, however, atoms are made up of electrons, and electrons become probability waves whose properties can only be represented by using terms like 'eigen values' and 'wave functions' which cannot be translated into anything directly picturable by the human imagination.[5] In relativity theory, space–time itself becomes a finite, curved and indented field, within which energy and mass are convertible. The 'real world' becomes an almost totally mathematical construct, and its physical reality becomes quite unimaginable. One has, not two, but three worlds – the mathematical, depicting the deep structure of quantum reality; the physical, consisting of more-or-less stable atomic and molecular entities; and the phenomenal viewpoint of sensory experience.[6] How these viewpoints relate to one another is perhaps the greatest mystery of our understanding of reality, and many think it to be in principle insoluble by any presently available theory.[7]

CONSCIOUSNESS AS AN EVOLUTIONARY GOAL

I am not proposing to do what the finest scientific minds of the present day profess themselves unable to do. But I have suggested that these three viewpoints express different ways of knowing and relating to one complex multi-layered reality. Talk of probability waves and eigen values does not undermine the reality of atoms and molecules. Talk of molecules does not undermine the reality of consciously apprehended beauty and meaning. On the contrary, it is the phenomenal world that provides the concrete particularities from which the other two worlds are progressive abstractions. From a quasi-Platonic mathematical world alone, one could not deduce the existence of the sub-atomic particles that operate in accordance with one subset of mathematical principles. From a sub-atomic world alone, one could not deduce the existence of the experiences that give conscious beings access to a public world of action and purpose. From the mathematical through the physical to the phenomenal, each level of reality has a more complex and substantive reality, introducing new properties and relations. Far from the physical world giving a complete explanation of the phenomenal, it seems to provide a partial and essentially abstractive account of the general formal structures of the phenomenal world, and the physical world too needs further explanation in terms of deeper mathematical principles. Thus no one could deduce a smell from an account of molecules and brain-states, but one can say that particular smells are triggered by specific brain-states, and that one can give an illuminating account of their relationships by studying various brain-states. The brain is so structured that it enables phenomenal data to be ordered intelligibly, in accordance with discoverable principles which establish coherent correlations between various data, such as smell, taste, touch and sight.

An analogy that seems appropriate is that of a computer program. A set of binary codes is constructed which in itself, without any translation program, is quite meaningless. But when one has a method of translation into phenomenal data, so that one binary number translates into a specific visual array on a screen (say, a letter), then one can code any amount of information into strings of binary numbers, which will store that information for retrieval at an appropriate time. So the structure of the brain is, in itself, quite meaningless, just a complex arrangement of electrons. But add a method of translation into phenomenal data, and the brain can store huge amounts of information in electron-patterns whose sole function is to store and provide the phenomenal data when required. The function of the brain is to receive, store and activate sensorily provided information on demand.

Just as it would be incredible to suppose that a computer program happened to compose itself by chance, and then translated itself into words which form a dramatic novel, so it is incredible to suppose that the brain originates by chance, and then translates its electro-chemical states quite unpredictably into thoughts, feelings and sensations, which give every impression of purpose and value. It is intelligible, however, to believe that computer programs are designed precisely in order to store information in their physical structure, and are provided with translation devices to bring this information to some consciousness. So it is reasonable to believe that the brain is designed to receive and store information, which can be translated into phenomenal experiences by some means which is not contained in the physical states themselves (the mode of translation into phenomenal data cannot logically be contained within a set of purely physical states).

In short, the overwhelming probability is that the structure of the human brain is designed to provide phenomenal experiences. Such experiences are not accidental by-products of complicated physical structures. They are the very reason why physical structures of such complication and order exist. As Sir John Eccles writes, 'An appealing analogy, but no more than an analogy, is to regard the body and brain as a superb computer built by genetic coding, which has been created by the wonderful process of biological evolution . . . the Soul or Self is the programmer of the computer.'[8]

To continue with the sensory example we have considered, smells are not wholly surprising, unpredictable and arbitrary results of complicated material interactions. They are the very reason why brains, and their complex causal interaction with chemical molecules, have the structures they have. When one thinks about consciousness in this way, the impression of purposiveness in evolution becomes virtually overwhelming. The whole process becomes intelligible, is explained, when it is seen as ordered towards the genesis of the sorts of phenomenal experience of an objective physical order that only highly structured central nervous systems can provide.

Evolutionary theory supports the claim that phenomenal consciousness introduces new properties into reality which are able to change physical states so that conscious organisms are better able to survive. The conscious recognition of one's environment gives one an advantage in the struggle for life. That presupposes that conscious states are not mere by-products of a physical process, but can play a causal role in modifying physical processes. The organism that develops a rudimentary eye is able to adjust its behaviour in accordance with its new sensory input. The registering of a phenomenal state leads to a modification of behaviour, and so to new forms of purposive causality in the world. The theist does not see this as a chance consequence of the working out of blind laws of physics. For a theist, the increasingly complex structuring of physical systems is directed from the first to the emergence of phenomenal consciousness, which can apprehend, interpret, understand and shape the physical world so as to realise new forms of value which can be enjoyed and shared with other conscious beings.

Even for an orthodox Darwinian, the advent of consciousness and purposive agency has a high survival value, and thus introduces a new and important element into the explanatory scheme of evolution. Yet survival as such is not important, except insofar as it is a condition for realising values that are important, values of the shared creation and apprehension of beauty. For an extreme Darwinian, it must always be an odd mischance, the incredible result of a million small errors in replication, that the universe, in human beings, has begun to understand its own structure and adapt it purposively to create a more stable and unequivocally desirable environment. How incredible it is that out of so many mistakes has emerged an outcome of such value! How much more plausible it is to suppose that the whole emergent process is set up precisely so that the universe could come to generate communities of beings capable of self-knowledge and self-control.

THE SECOND LAW OF THERMODYNAMICS AND THE GOAL OF CREATION

The process is, of course, by no means complete as yet. If standard cosmological models are correct, the universe will, if closed, continue to exist for at least 100 thousand million years. It is quite possible that the evolutionary process may continue during that time. This planet will become uninhabitable in between 900 million and 1.5 thousand million years, due to the increased luminosity of the sun, which will eventually destroy the earth completely. But within that time humans may have devised means of interstellar travel which will enable them to continue to exist in space or in other star-systems. It is also possible that within that time new and more intelligent and powerful forms of life will come into existence, which will displace humans from their present pre-eminent position on earth. Predicting the far future is a hazardous enterprise, but some cosmologists, admittedly straying somewhat into the field of science fiction, have postulated that life may engulf the whole universe, using supercomputers to generate new and improved life-forms, and eventually giving rise to a wholly self-knowing and self-directing totality, the Omega Point.[9] This seems so speculative that not much can be reliably based upon it. But it does show that once one sees cosmic evolution as progressing towards increased complexity, intelligence and awareness, one is tempted to extrapolate to a far future in which maximal awareness and intelligence will exist. For such a view, the universe may be said to generate God, or something very like God, an omniscient and omnipotent being.

Theists, however, are not committed to the hypothesis that the universe will somehow turn into God. On the contrary, they are usually quite careful to say that finite creatures will always remain different in kind from the infinite reality of a creator God, so that the goal of creation cannot be to create a second God. From a theistic point of view, the goal must be the creation of those distinctive forms of value that can be found in evolving, temporal and diverse communities of conscious agents, which can share in creative action and experiential knowledge.

From such a viewpoint, there is no necessity that human beings should be the final stage of evolution, or indeed that they should have existed at all. What a theist might reasonably infer is that the material universe would generate communities of conscious agents – whatever they look like and whatever species they belong to – which can fulfil at least a good deal of their positive potential by conscious knowledge of and co-operative interaction with God. Human beings certainly belong to the class of such agents, and so humans do represent one of the possible goals of creation. Humans equally certainly do not fulfil much of their positive potential – being largely destructive in nature – and lack to a lamentable degree any knowledge of or co-operation with God. One might well think that other life-forms may realise much more potential for creative action and extensive understanding, and co-operate much more fully and consciously with the unfolding of many divine purposes.

There is reason for saying, therefore, that there is more than one goal in creation. There is not just one Omega Point, or final state, at which some one goal for all creation can be realised. There are many goals, to be realised at many points in the created processes that form this space–time. The human species has not fully attained the goals proper to it, because of its self-entrapment in egoism and destructive desire. Yet there is a proper sort of fulfilment for human life, and if it can yet be achieved, that will achieve one goal of creation, even though there may be many others yet to be achieved in the cosmic process as a whole.

Even if one thought, as the cosmologist Frank Tipler does, that there is one final Omega state which somehow fulfils all goals of creation at once, that would not make the universe wholly self-sufficient or wholly autonomous. It will always remain dependent for its existence on that reality beyond it, which alone contains the power of being in itself, the being of God. So its final fulfilment will lie not simply in the realisation of a wholly self-possessed life. It will lie in its full and conscious relationship to the one source of all being that first formed its character and always sustained its development. The sharing in creativity and sensitivity that is the consummation of a worthwhile existence is itself brought to its greatest fullness by the sharing of the created order in the ultimate creative power and wholly sensitive knowledge of God, who infinitely transcends this, and every other, space–time.

That goal can be attained, however, without any reference to such a scientifically insecure postulate as an Omega Point. All one has to say is that the process of cosmic evolution is such that it will generate creatures that can share, consciously and fully, in the creative activity of God. Whether it exists for ever or for a finite period, this space–time will have realised forms of value which could not otherwise have existed, and which contribute to the divine being itself particular actualisations of its infinite potential for the unlimited expression of goodness.

It is often pointed out that, according to the second law of thermodynamics, the universe will, in the end, run down. All its energy will be dissipated and it will come to an end. Of course, it may be that some new state of affairs will result. Cosmologists have supposed that the universe might go into reverse and run through with all its physical laws in reverse order. Or perhaps a new set of laws will come into being, by some sort of quantum fluctuation. However complicated these possibilities may be, it does seem that the physical order as we know it, which supports forms of life like human beings, will not continue for ever. Whatever may happen to different orders of reality, this space–time and its laws will cease to exist. This may seem to be a difficulty for the claim that there is an objective goal in the evolutionary process.

Far from this being the case, it is exactly what might be expected if the theistic hypothesis is true. For a theist, this universe must be created with a purpose. That purpose, however, is not just the existence of some valuable final state – if it were, that state could just have been created at once, without going to all the trouble of evolution. The purpose must lie in the process itself, in the development from a specific set of physical potentialities of forms of conscious agency distinctive to this universe. The purpose is *growth* in creativity, sensitivity and community, a growth which involves the actions and responses of many beings generated within the universe, coming to actualise their own specific futures from the array of possibilities open to them.

Such growth will naturally issue in the realisation of desirable states, desirable both in themselves and because of the creative process by which they came to be. There is no reason at all, however, why such states should endure for ever. The theist will believe that they are held everlastingly in the 'memory' of God, but one might expect the creative process to continue with new projects thereafter.[10]

If the goal of evolution is the attainment of a sharing in the creativity, knowledge and bliss of God, it may thus seem that the attainment of such a goal within the cosmos will only be part of a continuation in ever-new forms of relationship, which are never finally completed. In other words, when the goals of the cosmos are achieved, the conscious agents the universe has generated will pass on to new forms of action and experience, beyond the confines of this space–time. From this perspective, the cosmic purpose is not to achieve a physical state that will never cease to exist. All physical states come to be and pass away. The purpose is to realise a specific set of values, through an emergent creative and relatively autonomous process. When that purpose is achieved, the cosmos can pass away. The values it realised will be conserved for ever in the mind of God, and the conscious agents it has generated can pass beyond its confines to new forms of creative and communal life.

Thus the second law of thermodynamics does not throw doubt on the hypothesis that the cosmos has a definite purpose, which will, by the power of God, be realised. God infinitely transcends this space–time, and the destiny of conscious beings is, above all, to be related to God in knowledge and love. One may look, not only for a full realisation of the distinctive values of this cosmos in future, but also for a continuation of personal lives beyond the confines of this universe, related to God in new forms of existence with new possibilities for creative growth and experience. That is a hope that goes well beyond the findings of the sciences at the present time. But is a hope that is not inconsistent with the scientific understanding of the universe, and that a belief in the ultimate intelligibility of the universe may even suggest. For a fully intelligible universe which aims at the realisation of personal potential in relation to God may well be thought to allow for such realisation beyond the confines of the physical realm which, while it generates persons, also often confines and limits their development. Such a thought finds a corroboration in religious experience and thought. If science shows the mind of God, religion shows the heart of God. Together, they provide an exciting and creative way to understand the ultimate nature of this universe, and even something of its infinite source and goal.

NOTES

[1] Richard Dawkins, *The Selfish Gene*, p. 63.
[2] There is a vast literature on this issue. One major work by a Nobel prize-winning neuro-physiologist which broadly supports the view I am taking is Sir John Eccles, *Evolution of the Brain: Creation of the Self*. Among philosophers, a spirited defence of the distinction of conscious states from physical states is given in Richard Swinburne, *The Evolution of the Soul*. My own view, slightly different, is sketched in *Defending the Soul*.
[3] See R. Taylor, *Action and Purpose*.
[4] Bishop George Berkeley, *A Treatise Concerning the Principles of Human Knowledge*.
[5] See J. Polkinghorne, *The Particle Play*.
[6] This is a slightly different classification from the well-known 'three worlds' of Sir Karl Popper. (See K. R. Popper, *The Open Universe*.)

He does not distinguish the mathematical from the physical, so his 'world 3', the cultural world, would be a fourth world in my list. I discuss it later, in the section on 'The Future of Evolution', pp. 167ff.

[7] Roger Penrose, *Shadows of the Mind*, ch. 8.7.

[8] Eccles, *Evolution of the Brain: Creation of the Self*, p. 238.

[9] See John Barrow and Frank Tipler, *The Anthropic Cosmological Principle* and F. Tipler, *The Physics of Immortality*.

[10] One form (not the only one) of such a view is found in the 'process thought' of A. N. Whitehead; see *Process and Reality*.

CHAPTER NINE

The Future of Evolution

DO MEMES TELL THE TRUTH?

There is one further step (the seventh stage) that the evolutionary process has so far taken on this planet. Not only have brains evolved, which make it possible for physical organisms to be self-aware and self-modifying to some degree, but, to quote Dawkins, 'a new kind of replicator has recently emerged on this very planet'.[1] Human animals group together to form cultures, which manage to break the first law of genetics (that acquired characteristics cannot be inherited). They do pass on acquired characteristics in a cumulative way. They do so by the communication of information which is learned and accumulated from one generation to another. So 'cultural transmission . . . can give rise to a form of evolution'.[2] Dawkins even invents a word for these new entities which 'propagate themselves . . . by leaping from brain to brain'.[3] He calls them 'memes', bits of information which replicate themselves in a new way, by being passed on in a culture.

Dawkins seems to realise that at this point one at last has a phenomenon which Darwinian natural selection cannot account for. Ideas may compete for space in brains, and some are more successful than others, they 'catch on' more effectively. But it hardly seems appropriate to talk of a physical mechanism for mutating ideas, or to speak of their propagation as a blind and purposeless one. For at this point awkward considerations of the truth of ideas arise. If we think we believe something because it is true, for a Darwinian this must be a piece of self-deceit, concealing the plain fact that we believe it because the idea has eliminated competing ideas from our minds, and is simply more successful in evolutionary terms.

But what could make it more successful? Unable to resist caricaturing religion, Dawkins suggests that 'The idea of hell fire is, quite simply, self-perpetuating because of its own deep psychological impact.'[4] That idea leaps from brain to brain, and succeeds because it appeals to some psychological need. If that is true, however, then it is pointless to try to criticise the idea of hell rationally, or to ask whether there is evidence for it or good reasons to believe in it, or whether it is true that hell exists or not. While the need exists, the idea will replicate successfully. The need might be eradicated, but not by argument.

The same will go for scientific ideas, too. Dawkins has a belief that scientific beliefs are true because they are supported by evidence. But on his own theory, one will only accept that 'beliefs are true if supported by evidence' if that is a psychologically appealing view. We should push the question further, and ask, is his own theory true? All he can consistently say is that the theory will perpetuate itself if it appeals to some deep psychological need. In other words, he is in exactly the same boat as the believer in hell. These are ideas that replicate because of psychological needs, not because of rational discussions.

But is *this* true? Do ideas replicate because of psychological needs? The belief 'Ideas replicate because of psychological needs' will only be accepted (replicate) if it meets some psychological need. If it does, then I will say it is true. If it does not, I will say it is false. But it is obvious that a belief may meet a psychological need if it is not supported by evidence (Dawkins thinks that belief in hell is like that). So the Darwinian explanation of beliefs in terms of replicatory success or psychological efficacy undermines Dawkins' own belief that beliefs should be supported by evidence.

Moreover, suppose I say that X is true if it meets a need.

I can always ask if *this* is true, if there is some identifiable need that X meets. But, on the theory, to ask if it is true is just to ask if it meets a need. In other words, I will be asking, 'Does "there is some need that X meets" meet a need?' But in order to find that out, I will have to look and see. Whereas, on the Darwinian explanation, I will just find myself believing it, and looking to see is irrelevant to the whole process.

For a consistent Darwinian it is clear that there is no point in appealing to evidence, reasons or universal agreement. One can only say that these are the ideas that inhabit one's brain. The very idea of truth becomes superfluous. That is what Dawkins is driven to. Unfortunately, it will not do. It is a self-refuting position, as many philosophers have pointed out. Even to say, 'The very idea of truth becomes superfluous' is to make a claim to truth. We cannot either say, 'What I am saying is not true', or 'What I am saying is true, but that means no more than that I am saying it'. For to say something, to assert it, is to say that it is the case, that it is true. It is not possible to speak and at the same time to eschew the idea of truth. In which case, there is something wrong with Dawkins' whole approach, and we have to say bluntly that we cannot consider ideas solely in terms of their memetic success. We have to consider their truth, and that may have little to do with whether they succeed in being passed on or not.

THE TRUTH-GAME

Proponents of the theory of natural selection have in recent years invented a new discipline known sometimes as 'sociobiology' and sometimes as 'evolutionary psychology'. It attempts to explain the survival and spread of human beliefs and theories in Darwinian terms. Its fundamental principle is that 'the ways in which we think . . . are themselves reflective of the ever-present pressure towards reproductive efficiency'.

Leading proponents of sociobiology are E. O. Wilson and Michael Ruse, Professor of Philosophy and Zoology at the University of Guelph. In his book, *Evolutionary Naturalism*, Professor Ruse gives a sophisticated defence of the theory that beliefs are selected by their survival value, and are consequently thought to be true.[5] However they arose, certain beliefs are advantageous in the struggle for survival. As Michael Ruse puts it, 'Those proto-humans who believed in $2 + 2 = 4$, rather than $2 + 2 = 5$, survived and reproduced, and those who did not, did not.'[6]

This idea works well at relatively simple levels. Obviously, organisms that form beliefs based on accurate perceptions of the environment will survive better than organisms that form inaccurate beliefs. Accurate beliefs are conducive to survival; and, if they, or at least the dispositions to form them, are genetically induced, they will be preferentially propagated. But can the propagation of all beliefs be explained in terms of their survival value?

Suppose – to take an example that might annoy Dawkins – belief in God aids survival. Perhaps it does so by making proto-humans think they should produce more offspring, by making them (mistakenly) hopeful about a joyful afterlife, or by making them ready to endure hardship for the sake of an imagined future happiness. Then, on the theory, belief in God will be genetically propagated. Such belief will spread rapidly throughout human populations. That is, after a relatively short time, all humans will think it is true that God exists. It is people like Dawkins who are the new genetic mutations which will spread only if they have superior survival value. And who knows whether that will be so?

Of course, Dawkins is not prepared to wait and see. He wants to say that belief in God is mistaken, that it is a false belief. He would say, in Michael Ruse's phrase, that 'we are deluded by our biology' into such things as theistic beliefs and beliefs in objective necessities in the world.[7] But if one claims that one's own belief in God is a delusion, then one is believing two contradictory things: first, that one believes it is true that there is a God, and second, that it is false that there is a God, since the first belief is a delusion. One must give up one of these contradictory beliefs. But if we give up the second belief, then we do not think we are deluded by biology. And if *that* belief is true, then we are not deluded by biology. If, on the other hand, we give up the first belief, then one cannot be deluded by biology, since we are not deluded. In either case, then, one cannot be deluded by biology.

So it must be other people, not us or not people in general, who are deluded by their biology. But how do we know they are deluded? Simply because they have beliefs that contradict ours. We account for this by saying that their genes cause them to have such beliefs. But we can now see that this is mere prejudice. We must in consistency admit that our genes cause us to have our beliefs, too. We are back to square one. The situation is just that genes cause us to have differing beliefs, and we will have to wait and see which ones help us to reproduce better. Those, whichever they are, will turn out to be the 'true' beliefs in the end.

At which point, it becomes rather obvious that the question of truth is simply a different question from the question of what survives better. It is in general true that true beliefs help survival (p entails q). But it would be a classical logical error (the fallacy of 'affirming the consequent') to argue that therefore beliefs that help survival must be true (that q entails p). Some beliefs may help survival but be false (belief that God will help the suffering is such a belief, according to Dawkins). Some beliefs may be true, but lower the chances of survival (belief that there is too much suffering to make life worthwhile).

When I seriously want to discover whether a belief is true, I would be wise not to consider at all the question of whether it is conducive to survival. It is most characteristic of the human mind to want to understand how things are simply for the sake of that understanding itself. If it came to an ultimate choice, many would prefer, and many martyrs have preferred, truth to survival. After all, when consciously considered, survival is not in itself an overriding value, whereas many (including Dawkins?) would think that the preservation of truth, for example in science, is.

Anyone who accepts evolution would agree that many simple beliefs have been built up because of their success in enabling organisms to cope with the environment. Natural selection does go some way to explaining the intellectual abilities of humans (though it does not explain how they developed in the first place). But it seems that there is a time in the evolutionary process when some conscious animal must have asked itself, for the first time, whether some belief was true, whether or not it had been hard-wired into the brain in some genetic way. There was a moment when considerations of intrinsic values and of freedom to pursue or ignore them entered into the evolutionary process, and from then on transformed it. That is perhaps the most significant phase change in the whole evolution of life on earth, the change from instinctive behaviour to behaviour that manifests a conscious response to value. That is what marks the enormous valuational difference between organisms and persons.

If this is the case, then the whole process up to that point can most intelligibly be seen as a teleological process, directed towards the emergence precisely of such truth-oriented beings, which are able to understand the nature of the universe of which they are part, and thus bring the universe to self-consciousness. The 'natural selection' of intellectual capacities, properly understood, turns out to be part of a progressive or purposive plan. Yet this is just what Darwinians like Richard Dawkins and Michael Ruse deny:

'Progress is impossible in the world of Darwinism, simply because everything is relativized in the sense that success is the only thing that counts.'[8] And again: 'One thing is absolutely fundamental: there is no progress.'[9] If truth is different from success, and if truth is an intrinsic value that cannot be measured solely by its conduciveness to survival, then such hard-line Darwinism must be at least mitigated by admitting the idea of progress. Dawkins and Ruse only oppose such a view because they are right in thinking that, once progress gains admittance, God is not too far behind. Their own relentless commitment to truth at any cost is perhaps the best argument for thinking that a fully coherent evolutionary view will turn out to be a strong support for a thoughtful theism.

THE WILL TO POWER: SOCIAL DARWINISM

Just as the idea of truth disappears in the extreme Darwinian view, so the idea of goodness and morality disappears, too. Darwin's talk of a battle for life, and his emphasis on the survival of the fittest, led historically to the development of forms of social Darwinism by Herbert Spencer and others. This development was usually formally fallacious, as the philosopher G. E. Moore pointed out. Moore's 'naturalistic fallacy' (which is almost always confused with a rather similar fallacy, the deontic fallacy of inferring 'ought' from 'is', exposed by David Hume) is the fallacy of identifying 'natural' or physical properties, like being the fittest, or strongest, or best at producing offspring, with moral properties, like being good or intrinsically worthwhile.[10] It is quite common to hear Darwinians say that animals 'co-operate for the biological ends of survival and reproduction',[11] or in order to pass on a part of themselves to future generations. Strictly speaking, of course, there are no ends in animal life for a Darwinian. They should really only say that animals happen to pass on their genes, some more than others. But having got started on 'purpose' language, they sometimes go on to say that this is what it is reasonable to do, or even what animals – and therefore humans – *ought* to do.

So Herbert Spencer argued that we ought to breed selectively for strength and intelligence, so that the strong will survive and the weak are gradually eliminated. Such views usually had racist overtones, so that the Aryan race was to become strong and outbreed the other races of the world. Social Darwinism is the view that the strong should be encouraged to compete and survive, and the weak must die. Thus the way was prepared, with the help of a dire economic crisis, for the transformation of German culture in one generation from the most intellectually active in Europe to being the home of a barbaric philosophy of 'blood and soil', of racial superiority and the will to power. It has often been asked how barbarity could triumph so quickly in such a high culture. The transformation is all too easy, when the high culture, seeing itself to begin with as heroically seeking truth at all costs, comes to regard human life as a purposeless mistake in a hostile universe of conflicting powers, which is indifferent to morality and sentiment. In a transforming instant, the highest intellectual truth is seen to be that intellectual truth is of no account. Reason destroys itself with the final reflection that it is an irrelevant scum on the blind and indifferent forces that drive life to inevitable destruction. Then the intellectuals, disillusioned with the life of a mind that has proved not even to exist, collaborate with the other forces of disillusion and help to rationalise (for that is their function, now truly perceived for the first time) the impulses of competition and greed constituting the naked will to power.

It is easy enough to point out the two fallacies underlying such arguments. The fallacy of reading a purpose into a process said to have no purposes, and the fallacy of inferring that, without need of further justifying reasons, one ought to further that purpose, are clear enough. Like T. H. Huxley, one can simply say that, if the process of evolution is bloody and cruel, one ought to oppose it, and do what is good for its own sake.[12] After all, as I have argued, we can agree on a set of objective basic values, things that all rational beings have a good reason to want. So we know what is good or rationally desirable. We ought to try to maximise those things. Morality does not depend on our acceptance or rejection of Darwinism, either as biology or as metaphysics.

But this is perhaps too facile. Views of what one ought to do (of whether, indeed, one ought to do anything) and views of human nature are more closely connected than that. Why should we aim at what is objectively good, especially when it is difficult or unpleasant, requiring self-sacrifice and an extension of our natural concern for others?

The theist has an answer to this question. The universe was created for a purpose, and this purpose is to realise intrinsic values among persons in relationship and community. Further, this is the purpose of a supremely good (worthwhile and desirable) being, whom it is our deepest fulfilment to know and love. I should aim at goodness because that is the objective purpose of my existence. This purpose, for the theist, is real and rooted in an objective cosmic mind. It is not just a projection of human motives on to a neutral universe. I should follow that purpose because I can see that it is a good purpose which can and will be realised, and also because in doing so I will have a good chance of realising one of the greatest human goods, which is an obedient relationship of love with God.

The Darwinian, by contrast, must say that there is no cosmic purpose likely to be realised. There is no cosmic mind to be related to in obedient love. There is life, struggle, copulation and death, and that is all. I can still aim at goodness for its own sake. But I now see that the very idea of 'goodness' is constructed by the mind as a survival strategy, so that animals will act in ways conducive to the survival of a set of genes. To aim at goodness for its own sake now becomes a sort of pathology. It is to be deceived by a relatively stable evolutionary strategy, to mistake the means for the end. The real reason for moral beliefs, the reason they have got implanted into human minds, is that they are conducive to the propagation of more genes. If I think the reason for moral thought and action is to realise intrinsically worthwhile states, I have been duped.

Can I see this, and still simply aim at goodness and virtue? It will hardly be rational to do so (though admittedly reason is no longer very important on the Darwinian view). Once the sociobiologists have enabled me to see through the evolutionary strategy of morality, I can be liberated to pursue the real goal of morality more directly, and devote myself to propagating my genes as prolifically as possible. Of course, I now have to remind myself that I am still being duped, as there is no 'real goal' at all. The pursuit of goodness has become pathological, and there is absolutely nothing that should or can take its place. Morality has been dissolved. It has simply ceased to be.

There are those who can embrace this as a sort of liberation from convention and hypocrisy. But when one sees that theism is able to give the pursuit of goodness a strong rational basis, and that social Darwinism is going to demolish notions of human dignity, freedom and autonomy along with morality, rationality and truth, one may well pause for thought. Of course one cannot reject Darwinism just because it has unpleasant social consequences. But if one can see that it undermines some of the most important and strongly held human values, one will enquire very carefully into its credentials. Are we not more certain of some fundamental values, of the importance of morality, than we are of the metaphysically inflated worldview of social Darwinism? Does practical commitment not take precedence over theoretical speculation, and indeed provide the real basis for much speculation in these very fundamental areas? I think it is clear that it does, and that we should have the courage to seek a worldview that can give a rational underpinning to our deepest practical and evaluative beliefs. It is one of the strongest arguments for theism that it can do that. It is a strong argument against the Darwinian worldview that it undermines our deepest evaluations, while it seems based on evaluative arguments itself. The evaluation it is based on is that human lives are torn by contrary desires and strong aversions. Our world is a theatre of conflict, a ceaseless struggle for dominance and submission, a desperate struggle for existence in a world of scarce resources, of few lasting pleasures and innumerable dangers, difficulties and accidents of fate. As Dawkins pessimistically writes: 'If there is ever a time of plenty, this very fact will automatically lead to an increase in population until the natural state of starvation and misery is restored.'[13] In such a world, the natural thing is to steal a little transient pleasure from indifferent chance, before surrendering to the inevitable futility of death, and to ensure that one's own kin starve last of all. Our genes will take care of that, anyway, and it is natural to let them have their way.

Beneath the allegedly neutral descriptions of Darwinism lies a set of highly evaluative judgements about human nature and its proper forms of activity. Put at its most stark, the choice is between seeing authentic human life as a life of participation in a supreme reality of wisdom, compassion and bliss, or as the triumph of the will to power and survival, a temporary triumph to be sure, doomed to final failure. At this point, even if the theoretical arguments on both sides were equally balanced – which I have argued they are not – a sense of moral commitment and hope for the future might well incline one to bet one's life on the theistic view. Even if one was wrong, the world might be better for it.

More importantly, just as one cannot avoid making a claim to truth, even while claiming that truth is a device for genetic dominance, so one cannot avoid making a moral commitment, even while claiming that morality is a survival mechanism. Moral commitments are unavoidable, and humans seek to make such commitments as are based on good reasons. Darwinism either deprives morality of any rational basis at all, or it seeks to absolve itself of all moral responsibility by asserting that our actions are determined by our genes (which, unfortunately, are inherently selfish). An alternative view is that it is precisely in moral consciousness, in the sense of obligation and justice, that one senses the transcendence of personhood over biological necessity. Only a theory that is completely certain should be allowed to undermine this moral sense. Metaphysical Darwinism is far from being such a theory. Indeed, its inability to account for the moral consciousness in a satisfactory way is one of the strongest arguments for its incompleteness as a total explanation of human behaviour, and therefore of the evolution of life.

EVOLUTIONARY ETHICS

A sophisticated defence of evolutionary ethics has been given by Michael Ruse. But his defence does not succeed in overcoming the problems mentioned in the previous section. In many ways, it highlights those problems even more starkly. Professor Ruse begins by showing that 'altruistic' behaviour can lead to the 'selfish' end of replicating the genes of the agent more effectively than 'selfish' behaviour might do. It is therefore possible to give an account of moral altruism by appeal simply to the principle of natural selection. One simple example would be the way in which a bird may give its life for the survival of its chicks, or even of the chicks of its neighbours. By such apparently self-sacrificing behaviour, more of its genes survive than would do if it flew away, leaving the chicks to be eaten by some predator.

While this is quite true, Ruse himself notes that the terms 'altruism' and 'selfishness' are being used in a technical way. Altruism in this sense is not rationally considered self-renouncing action. It is genetically determined behaviour, exhibited naturally and unthinkingly by organisms. There is no intended goal of such action. It is just that organisms genetically programmed to behave in such ways will ensure the maximal replication of the genes that programme such behaviour, which are present in their kin. Thus 'altruistic' behaviour will be propagated throughout the kinship group, and will be biologically determined in all its members.

Ruse says that 'Morality is no more . . . than an adaptation, and as such has the same status as such things as teeth and eyes and noses.'[14] That may well be true at the level of sub-rational behaviour. Animals may have no more choice as to whether they should act 'altruistically' or not than they have as to whether they shall have noses or not. But what happens when choice arises? One standard response is to deny that choice ever really arises at all. But strong biological determinism flies in the face of experience. It certainly does not help anyone to answer the question of what they ought to do, since all it says is that you will do whatever your genes have programmed you to do. The fact is, however, that the question, 'What ought I to do?' does arise for humans, and it demands a reasoned answer. But how can a biological account provide anyone with reasons for acting, since all it does is to provide causes?

'Sometimes when one has given a causal analysis of why someone believes something, one has shown that the call for reasoned justification is inappropriate,' writes Ruse.[15] That is, indeed, sometimes true. But if I want to know whether I ought to give money to charity, it is of no help to be told that asking for reasons is inappropriate. I may have some inclination to give to charity, but I want to know why I should. Or at least I want to know whether it is true that I ought to do so, and whether I should take that 'ought' seriously.

Precisely that question is ruled out by the evolutionary ethicist. For 'evolution has given us this logically odd sense of oughtness'.[16] It has stamped in us the belief that there is some sort of objective truth about what we ought to do. Yet this is 'a collective illusion of the genes'.[17] Ruse is now, or should be, in a quandary. If evolution has hard-wired into us a belief that there are objective moral obligations, then we will believe that there are. But if we see that such belief is 'an illusion of the genes', then we will believe that there really are no such obligations. Presumably, then, we could adjust our sense of obligation if we so wished – in order, perhaps, to aim more rationally at long-term self-interest. If 'morality is a creation of the genes',[18] we are free to modify it if we can, if necessary by chemical adjustment of the relevant bits of DNA. Are we free to modify our moral sense by rational reflection and conscious goal-setting or not?

Ruse says both yes and no. On the one hand he says, 'We are not free to decide whether or not murder is wrong.'[19] That belief is biologically imprinted, and 'your genes are a lot stronger than my words'.[20] In that mood Ruse says, 'There is no possible way in which knowledge gleaned from an intellectual theory as such could change my thinking about right and wrong.'[21] We are just programmed to think we ought to do certain things, and that is that. On the other hand, he also asks, 'Could we possibly owe it to our children to be immoral?'[22] And he suggests that 'A better understanding of biology might incline us to go against morality,'[23] for example to break the moral rule against killing in a particular case, for the sake of the long-term goal of preserving human life in general.

He really cannot have it both ways. Either I will be genetically programmed to think that there is an objective and binding morality – in which case I must believe that the evolutionary analysis of ethics simply in terms of a non-purposive and randomly generated survival mechanism must be false; or I am able to see that obligations are illusions of the genes and to modify them for the sake of rational goals – in which case the evolutionary analysis is again false, since 'morality' in its non-illusory sense is now seen to be, not a creation of the genes at all, but to be based on the rational construction of intended goals of action. Either way, the 'evolutionary ethics' analysis does not do well. It comes out as false on both possible interpretations.

Suppose, however, that one adopts a modified view of the analysis, so that one can account for the existence of some sense of obligation by reference to biologically imprinted beliefs, but then allow for its modification by rational reflection. If it is not possible to think of any objective or rational moral truth at all, yet one is able to modify or reject one's moral programming, who is to say how it should be done? It may seem that one should aim at more perfect forms of positive social interaction.[24] But since this 'should' must be evacuated of any moral force, why should one not rather aim exclusively at one's own, or one's family interests? Why should one not aim at the survival of the strong, and the ruthless elimination of the weak? Why should one not aim at almost anything, however selfish or brutal?

As Ishtiyaque Haji has argued,[25] it is always compatible with one's biological interests that one should behave from purely selfish motives. So selfish action would not contravene any biological necessities. In any case, once one begins to think about it, the survival of genes like one's own is not in itself a terrifically desirable goal. It may seem much more sensible to aim at one's own pleasure, whatever happens to one's genes. The conclusion seems to be that there is no rational argument from what has happened in biology to what one ought to do. Such arguments almost always commit the modalist fallacy (different from both Moore's naturalistic fallacy and David Hume's deontic or 'is-ought' fallacy) of inferring 'ought' from 'must'.

A good example of the modalist fallacy is found in the work of the sociobiologist Robert Richards, where he tries to defend a biological justification of morality. It goes like this: 'The logical movement of the justification is from – (a) the empirical evidence and theory of evolution, to (b) man's constitution as an altruist, to (c) identifying being an altruist with being moral, to (d) concluding that since men so constituted are moral, they morally ought to promote the community good.'[26] What is established in moves (a) to (c) is that 'being moral', in the sense of acting altruistically, can be shown to have had a survival advantage. Thus it is likely that moral action is genetically programmed into human beings. The principle of natural selection has brought it about that human beings are programmed to act socially. That is, they *necessarily* act in 'altruistic' ways. They are necessarily 'moral', in that they act according to principles of helping others.

Actually this is very far from being obvious. Francisco Ayala convincingly argues that 'some norms may not favour, and may hinder, the survival and reproduction of the individual and its genes'.[27] One such norm is that of celibacy among certain classes of human, which endures in some societies despite its counter-reproductive efficiency. Nevertheless, even if one allows the thesis that moral action will be selected preferentially, assertion (d) in Richards' argument is fallacious. It does not follow from the fact that humans necessarily act so as to help others, that humans morally ought to act in that way. Not only does it not follow, the alleged conclusion actually contradicts the premiss.

For if one necessarily acts in a certain way, then it is not possible for one to act in any other way. But if one morally ought to act in a certain way, then it must be possible for one to act in that way, and it must also be possible to fail to act in that way. That is, it must be possible for one to act in some other way than that specified in the obligation. Otherwise there is no meaning in saying that one ought so to act. It follows that the same act cannot be both necessary and morally obligatory. If humans necessarily act altruistically, then it cannot be the case that they morally ought to act altruistically. To confuse these modalities of necessity and obligation is the modalist fallacy.

The evolutionary ethicist thus seems to be left with no criterion of moral choice, and with no way of preserving a sense of the importance of morality at all. At this point, a theistic analysis possesses a decided advantage. It can see 'altruistic' behaviour as largely rooted in natural selection, but also see in the evolutionary process some important elements which Ruse rejects, the elements of progress, directionality and theism. Then one could say that God has set the evolutionary process up so that generally altruistic natural behaviour patterns would develop among groups of organisms. This would lay the basis for the emergence of a reliable knowledge of basic objective moral norms, which could be further refined by rational reflection – influenced by tacit or explicit knowledge of the divine being. On this account, knowledge of obligation would in fact be one form of a developing and conscious knowledge of God, of the ultimate truth of being. Its biological origins would be a natural consequence of the grounding of the whole evolutionary process in a divine plan. They would not in themselves be the ultimate foundation or truth of morality. Morality could then be seen to be both reasonably grounded in the evolved nature of human persons and to be oriented towards an increasingly conscious and reflective relationship with a transcendent reality and truth.

ESCAPING FROM THE TYRANNY OF THE GENES

Once one brings into play ideas of truth (of what is believed to be the case, whether or not it is psychologically appealing), of beauty (what is valued for its own sake, and not for its quantity, survival or replicatory efficiency), and of goodness (what is done because it is right to do it, not because it is conducive to survival), one has transcended the realm of purely physical causes and effects. These ideas lie at the very heart of the scientific enterprise. They fill Dawkins' own work with a passion that communicates itself unmistakably. Dawkins cares passionately about truth, about accepting beliefs, not because they are comforting, but because the evidence supports them. He cares passionately about the beauty, elegance and simplicity of the fundamental laws of nature. He cares passionately about goodness, especially about the virtue of intellectual honesty and the virtue of facing up to unpalatable facts. But why should he care about these things, which are traditionally called the transcendental values? It would seem demeaning to say that these are just memes which have leaped into his brain, and determine his beliefs and writings.

    'Consciousness', he writes, 'is the culmination of an evolutionary trend towards the emancipation of survival machines as executive decision-takers from their ultimate masters, the genes.'[28] Remarkably, after all he has said about evolution being blind and without purpose or goal, he now admits that there is an evolutionary trend, a direction in evolution. This trend is towards 'emancipation' and towards 'executive decision-making'. 'Our genes may instruct us to be selfish,' he says; 'But we are not necessarily compelled to obey them all our lives.'[29] Humans can be free from the dictates of physical causality, and this freedom has a clearly moral dimension to it: 'We alone, on earth, can rebel against the tyranny of the selfish replicators.'[30] The goal towards which evolution works is freedom to choose what is right, freedom from selfishness, freedom to be altruistic, to explore the universe, understand it and shape it to good. He ends *River out of Eden* with a quotation from Wordsworth, which sees in the admirable Newton 'A mind for ever Voyaging through strange seas of Thought, alone'.[31] This, for him, is a goal of intrinsic value. From the blind tyranny of purposeless causality, there arises a being that can conceive worthwhile purposes, liberate itself from the shackles of nature, and rise to the freedom and the ultimate good of the contemplation of the beauty of truth.

I believe this is the authentic voice of Dawkins, the voice of a man driven by a passion for goodness, for intellectual freedom and for truth at any cost. But the question that arises for his own interpretation of the evolutionary process is: why should one care about that? Why should we admire it, or be moved by it? We are, and we should be. But how can his own views justify it?

The obvious remedy is to accept that there is a 'trend', an inevitable tendency in evolution, to generate forms of consciousness motivated by goals of truth, beauty and goodness. But this is not a blind trend, exercising a tyranny of genetic selfishness. In fact, the very probabilistic nature of genetic change, which Dawkins sees as demonstrating lack of purpose, frees the process from tyranny. It allows an openness in nature, within which conscious freedom will be able to work. At the same time, there is an inexorable increase of complexity and integration, which makes consciousness possible. The best explanation of such a tendency is to say that it exists for the sake of the goal to which it tends, the existence of free consciousness. The replication of genes never was the purpose of evolution. Replication and directional

change were necessary conditions of the emergence of forms
of conscious life that could bring the material universe to
understand and shape its own being. Evolution is the process
by which the physical cosmos comes to generate beings that
understand themselves and control their own natures. From
the first moment of the Big Bang, it has been directed towards
generating forms of consciousness that could orient
themselves to the pursuit of truth, beauty and goodness. Its
goal is the existence of a universe that generates beings who
understand how to generate worthwhile states creatively, and
who find fulfilment in contemplating them. In that respect,
such beings are images of the divine being, which finds
supreme happiness in the contemplation of the infinite
perfections of its own nature.

THE EIGHTH STAGE

If the whole cosmic process is essentially and inevitably
directed towards producing fully self-knowing and self-
controlling beings, the obvious explanation of this is that
some mind must already exist to give it that direction, and
oversee the process. In other words, God, as a knowing
agent, must already exist, apart from the universe, to direct
it towards its goal. The goal must, after all, be consciously
envisaged and events must be wisely ordered to produce it –
both of which require a vast intelligence already in
existence.

The cosmos does not contain infinite perfections. It is a
finite space–time complex, and it realises a specific set of
possibilities out of an infinitely larger array. This universe
traces just one path, by which a material cosmos can generate
conscious agency. Its existence will end, at which point either
the story will be complete, having achieved its goal, or it may
pass, as most theists think, into the wider reality of the supra-
cosmic God. But one can see what is meant by saying that the

universe is, in Plato's words, 'a moving image of eternity'. The cosmos is an image of God insofar as it comes to embody self-awareness and free agency.

This suggests another important corollary. If the goal of the cosmos is to generate conscious agency, and if its creator is the supremely conscious agency of God, then the realisation of the goal must include a knowledge of the existence of God, of the dependence of the cosmos upon God, and of the likeness of mental existence to divine existence. In other words, the goal is likely to lie in the realisation of a conscious relationship of the cosmos to its creator. The cosmos will be an integrated community of communities of self-aware and self-directing persons in relation, ordered by a conscious and harmonious relationship with the infinite personal source of all beings. This is the future eighth stage of the evolution of life. Just as nuclear particles point beyond themselves to the existence of self-organising molecules, and molecules point beyond themselves to the existence of consciousness, so consciousness points beyond itself to the existence of an infinite consciousness, in relation to which its goal is achieved. The whole physical cosmos possesses the property of self-transcendence, of orientation to a goal beyond itself, in relation to which it achieves its proper fulfilment. In this process, human beings seem to stand at a crucial transitional point. At this point, at least on earth, the goal comes to consciousness for the first time, and the cosmos comes consciously to relate itself, through human lives, to its infinite source. The world's religions are paths by which God discloses some aspect of the divine reality to human beings, and by which humans come to relate themselves in conscious response to God.

Dawkins sees religion as a blind faith in primitive myths from a superseded past. Thus it is bound to conflict with science, which requires critical questioning, empirical investigation and the replacement of old theories by new.

Unfortunately, Dawkins' own presentation of Darwinism propounds a dogmatic, materialist and anti-purposive theory which goes well beyond the evidence, and discourages serious enquiry into the place of religion and morality in human life. My suggestion is that most of the evidence from evolutionary biology that Dawkins so brilliantly presents actually points in the direction of purpose and a built-in tendency towards the realisation of value in the cosmos. If religion is a positive response to intimations of purpose, of truth, beauty and goodness in the universe, and a pursuit of those things by a self-transforming acceptance of their magisterial authority, then it is not a defensive retreat into blind faith, but a committed advance to a wider vision of human existence in the cosmos. The theory of evolution is not in opposition to religious claims. On the contrary, it provides a new and exciting vision of the way in which the purposes of a divine creator are worked out in the cosmos.

NOTES

[1] Richard Dawkins, *The Selfish Gene*, p. 206.
[2] Ibid., p. 203.
[3] Ibid., p. 206.
[4] Ibid., p. 212.
[5] Michael Ruse, *Evolutionary Naturalism*, p. 157.
[6] Ibid., p. 163.
[7] Ibid., p. 183.
[8] Ibid., p. 178.
[9] Ibid., p. 185.
[10] G. E. Moore, *Principia Ethica*, p. 9.
[11] Ruse, *Evolutionary Naturalism*, p. 238.
[12] T. H. Huxley, *Evolution and Ethics*.
[13] Dawkins, *River out of Eden*, p. 132.
[14] Ruse, *Evolutionary Naturalism*, p. 241.
[15] Ibid., p. 249.
[16] Ibid., p. 245.
[17] Ibid., p. 250.
[18] Ibid., p. 290.
[19] Ibid., p. 253.
[20] Ibid., p. 256.

[21] Ibid., p. 284.
[22] Ibid., p. 282.
[23] Ibid., p. 283.
[24] This is the argument advanced by J. Collier and M. Stingl, in 'Evolutionary Naturalism and the Objectivity of Morality'.
[25] I. Haji, 'Evolution, Altruism and the Prisoner's Dilemma'.
[26] R. J. Richards, *Darwin and the Emergence of Evolutionary Theories of Mind and Behaviour*, p. 289.
[27] F. Ayala, 'The Biological Roots of Morality', p. 237.
[28] Dawkins, *The Selfish Gene*, p. 63.
[29] Ibid., p. 3.
[30] Ibid., p. 205.
[31] Dawkins, *River out of Eden*, p. 161. From Wordsworth, *The Prelude*, Book 3, 1850.

CHAPTER TEN

Suffering and Goodness

IS EVOLUTION CRUEL?

I have claimed that God is the best explanation of the evolutionary process, both in the cosmos as a whole and in its biological form on earth. The hypothesis of God is a simple and elegant one, and explains the cosmos as intended to realise the goal of generating a self-aware and self-directing physical reality, or a community of such realities, capable of conscious relation to their creator. God provides the basis of an all-embracing cosmic purposive explanation, and the existence of God itself is purposively explained by pointing to its actualisation of a maximal set of coexistent values. Such a God necessarily exists, as the ontological basis of the existence of all possibilities, of which God actualises the maximal compossible set in God's own being. It thus provides both the 'logical rigidity' or necessity that physicists desire,[1] and the self-explanatory supreme value required by purposive explanation. In this sense, God is a supremely simple integrating reality, and the best possible explanation for this, or indeed any other intelligible and value-actualising cosmos.

At this point, however, Dawkins produces his trump card. The universe just is not the sort of universe one would expect if there were such a God. A supremely perfect God would not create a universe like this, and so God does not render this universe highly probable after all. 'The universe we observe has precisely the properties we should expect if there is, at bottom, no design, no purpose, no evil and no good, nothing but blind, pitiless indifference.'[2]

Does the universe exhibit pitiless indifference to value, or is it essentially directed towards the free realisation of

truth, beauty and goodness? There is surely something to be
said on both sides. The universe is obviously not created by a
being whose purpose is to prohibit all suffering or to preserve
all sentient creatures from harm. Perhaps some pictures of
God have suggested that such would be God's purpose, but if
so they have clearly been totally out of touch with reality. On
the other hand, as I have consistently argued, evidence of
purpose is to be found almost everywhere in the universe. We
can make a good guess at what that purpose is – namely, the
generation of communities of free, self-aware, self-directing
sentient beings. There is not an indifference to value, though
value seems to be realised in painful and circuitous ways, and
at the cost of many individual lives. It seems to be
accompanied by the realisation of many disvalues, disvalues
of constraint, ignorance and failure.

What this suggests is that the universe is not blind and
purposeless. But it does not suggest that the universe has a
creator who can and will eliminate all suffering, whenever
and wherever it threatens to occur. It suggests that there exists
a creator who gives the universe a purpose which involves,
and does not exclude, such dysteleological (seemingly anti-
purposive) factors as suffering, frustration and death. It is a
moral purpose, in the sense that it is connected with the
realisation of the values of truth, beauty and goodness. But it
is not a purpose that can be actualised without suffering of
many kinds.

The question to ask of the theistic hypothesis must be: is
this purpose a worthwhile goal, and could it be a purpose
proposed by a God who is supremely perfect? If the answer to
the first part of this question is affirmative, and if the process
has to be in general like it is for the goal to be achieved, then
the second part of the question is answered in the affirmative,
also. For if the existence of suffering is a *necessary* condition
of the realisation of a worthwhile goal, then even a creator
God could not eliminate it, and choose that goal.

So we are left with the following question – is the goal of this universe, the development from a material mass–energy complex to the existence of communities of free, self-aware, self-directing beings capable of conscious relationship to the creator, a worthwhile goal? It seems obvious that it is. But we have to take into account the cost of achieving such a goal. What is that cost? Is it necessary? And is it justifiable?

THE COST OF CREATION: AN INTEGRAL UNIVERSE

It is modern physics that helps to show how the cost is necessary. The interconnection between the fundamental constants of nature is such that it could not be other than it is, without the whole structure falling apart and ceasing to exist.[3] If one is to have emergence in the universe, one has to have change and death, the elimination of old properties to make way for new. Out of the decay of nuclear particles, atomic structures are built. Out of the explosions of stars, carbon, the basic stuff of life, is formed. The pattern of the universe is that continual destruction gives birth to new life. Nothing is literally permanent, and yet each element depends upon all the others, in an integrated web of being which continually forges from itself new forms and levels of existence.

Human beings could not be simply removed from such an interconnected context and placed somewhere else, in a pain- and conflict-free environment. It may, religious believers usually think, be possible to remove them to such an environment later on, if certain conditions are fulfilled and if they undergo quite a transformation of nature. But they would not be human, they would not be conscious and free parts of this physical cosmos, if they were formed in some other cosmos. There could, no doubt, be different sentient beings – angels, perhaps? – which do live in different environments and have less troubled natures. But such beings would not be human. They would not be us. So, either we would not exist at all, or we have to exist in this very universe, with its general structural pattern of interaction, conflict, decay and renewal. Such a pattern exists at the nuclear level, at the level of evolutionary biology, and at the level of animal relationships. It is this pattern alone that makes *our* lives possible. Insofar as the physical sciences make clear our part in this integrated physical context, they show that the possibility, and at least some actuality, of conflict, suffering and failure cannot be eliminated from such a universe.

The best hypothesis seems to be that the cosmos has a goal of great intrinsic value, and that the pursuit of this goal has a necessary cost. This suggests that there is a cosmic mind which sets the goal and directs the universe towards it, in full awareness of the suffering necessarily involved in that process. Such a view of God may offend those who have a very sentimental view that a God should not permit any suffering at all. But what they are usually thinking is that God could vary the conditions of existence, and still achieve the same goal, or that perhaps God should not have this goal at all. When we see that the goal is internally related to the nature of the process itself, we see that the conditions of existence could not be varied. And when we see that, if God did not have this goal, we would not exist at all, we may think twice before blaming God for having this goal.

The one really difficult moral objection to God is that we may think it is better never to have been born. If we really do think this, in full knowledge of all the facts, then, and only then, we will conscientiously reject the divine purpose for creation. It is not, even so, that we will be able to deny God's existence, or that there is a goal in creation, or that much good will come of it. It is just that we will, in Ivan Karamazov's words, 'return our tickets', preferring total non-existence to existence in a universe like this.[4]

In Dostoevsky's story, Karamazov turns the screw by asking us to contemplate the suffering and death of an innocent child. Nothing, he suggests, not even an eternity of bliss, could justify the existence of such innocent suffering. It is as if we are asked to accept the suffering of the innocent as a means to our own happiness. That is something he cannot bring himself to do.

In seeking a theistic response to this case, it is vital that we do not underestimate the horror of suffering, that we do not come to regard it as something we can accept with equanimity and without protest. The suffering of innocent children is something that must be opposed and eliminated wherever possible. We can never accept it or do nothing about it, on the supposition that it will bring about some greater good. Innocent suffering is always evil, and must always be eliminated if possible.

It may seem as if God could eliminate suffering, for is God not all-powerful? But we have seen that even God cannot eliminate or change the necessary conditions of the cosmic goal. Among these conditions is the possibility of much suffering, and probably the actual existence of some suffering. Not all the suffering of our world is necessary, by any means. But the possibility of it may be necessary. That entails that, under certain conditions which are in principle avoidable, and which the creator may not intend or desire, that suffering the possibility of which is built into our world may be actualised. The fact is that, in a world in which many creatures are free within limits to determine their own futures, there exist many states that God neither intends nor desires. This may seem odd, but it can easily be shown.

God is able to bring into existence beings with the power of creative action and subjectively valued experience. God could terminate their existence at any time, or could

wholly determine their natures and actions. But if God wishes
to give them a limitedly autonomous power of action and
evaluation, God will refrain from wholly determining them.
In that case, if such creatures decide to do things by their own
free decision, God cannot be said to intend what they do.
God may (or may not) approve of what they do, but
intending entails bringing about, and if God has decided not
to bring about creaturely decisions, but to let creatures bring
them about themselves, then God cannot intend those
decisions. God obviously cannot bring it about that creatures
do what only they have decided to do, or they would not
really have made the decision. So, if there are truly free
creatures at all, there will be many occurrences – their own
freely intended acts – that God does not intend, even if all
they do agrees with what God wishes or even commands.[5]

THE CONSEQUENCES OF FREEDOM

Free creatures therefore can, and indeed must, do things that
God does not intend. But can they do things that God does
not even desire? Supposing that God desires only states that
are intrinsically good, how could a free creature desire a state
that is not intrinsically good? In the abstract, this may seem a
difficult question, but in practice nothing is more obvious.
What sorts of things do human beings tend to choose? They
choose such things as sensual pleasure, excitement,
competition, self-determination, courage, independence,
adventure, experiment and risk. In other words, they choose
goods that entail the risk of suffering, and that indeed entail
that some suffering will exist.

No rational being would choose a course it could
foresee would lead to endless pain – that is why the conscious
choice of hell seems irrational. But one may rationally choose
a sort of pleasure that carries a great risk of pain. For
instance, the thrill of motor-racing is not obtainable except by

actually careering around a track at very high speeds in a fragile box on wheels. The danger involved is part of that thrill. But where there is danger, accidents may happen, and pain or death will result. The racing driver does not choose the pain. But such a driver chooses a sort of good (the pleasure of dangerous sport) where pain might well result. If pain occurs, one need not approve of it, and one will certainly seek to avoid it. But one will have to accept that it was always a real possibility of a sort of goodness that was rationally chosen.

In general, the pleasures of selfish gratification carry a high risk of suffering. One can pursue such pleasures, thereby accepting the risk. One hopes to avoid suffering, or at least that one may pass through it in time. But it may be that one becomes habituated to selfish attachment, so that it seems virtually impossible to escape, even when it has indisputably led to personal suffering. It would take me too far from the present discussion to pursue this point in detail, but it does seem that a free creature may rationally choose a course of desire, conflict and attachment, which may well lead to personal suffering. One is not choosing the suffering, but one is choosing a course of life that leads to suffering, even though one hopes to escape it, either altogether or at least eventually.

Since one may assume that God does not desire creatures to suffer, such free creaturely choices will almost certainly lead to states of suffering that God does not desire. It is in that sense that many things in creation may be opposed both to God's intentions and to God's desires, even though God indirectly brings those things into being, through the agency of the free wills God directly brings into being. Thus much possible, though avoidable, suffering in our world can become actual despite the fact that God does not desire it to become actual.

The choice of sensual desire does not only involve

consequences for oneself, as though one lived in complete isolation. It will involve serious consequences for others, especially those to whom one is most directly related. Again, this is clear from everyday life. If I have an accident at high speed on the race-track, my car may plough into a group of spectators, killing and injuring many. They suffer, not for what they have done, but simply because they are in the same world as I am. Similarly, if I pursue a course of ruthlessly competitive ambition, for the pleasure of winning fame and fortune, my family may suffer neglect or even ill-treatment, when I am under pressure. They will suffer, even if they are wholly innocent, just because they have to live with me. The people whom I may use and manipulate in prosecuting my ambitions will also suffer. Selfishness is contagious and destructive in its effects, and the suffering it brings with it affects most of those to whom the selfish will is related.

In a world in which creatures are closely related to one another by bonds of affection and loyalty, suffering will spread widely through communities, because of the selfish acts of some of their members. The bitterness and frustration such suffering causes will in turn compound the destructive effects of selfish acts, as hatred and resentment multiply. Communities that could have been resources of strength and co-operation become distorted centres of hatred and fear. In such a world, rationality is obscured and, driven by passion, creatures even torture and kill one another.

In this light, one can see the suffering of an innocent child as a result of the corruption of the goods of freedom, desire and community. The consequences of the choice of self-interested desires are, ironically, the destruction of the very freedom and rationality that were the first conditions of such a choice. Creatures can become trapped in a world of irrational and self-seeking passions. In such a world, innocent children suffer, not as a means to a greater good, but as a consequence of free choices made in opposition to the desires of the creator.

CHOOSING LIFE

At this point, the religious believer will agree that, if this is the last word, the cost of such freedom, of such a denial of God's will, is too great, that a world of hatred, greed and ignorance is unbearable. A world dominated by hatred and fear of others, greed and the will to power, ignorance of the reality of God and the possibility of goodness, is a world without point or hope. To accept and enjoy existence in such a world is merely to steal a little transient pleasure from an indifferent universe, before surrendering to the inevitable indignity of death.

But for a theist, this is far from being the last word. The picture of a human existence without point or hope is the result of a deep ignorance, itself the product of immersion in self-centred desires. This is a picture that creatures have largely drawn for themselves, of a world estranged from God, pitiless and indifferent. From the viewpoint of such an estranged world, nature itself comes to seem cruel and uncaring, a breeding ground for millions of suffering, unhappy creatures, living and dying without point or purpose. When God dies, the Darwinian metaphor of a continual and purposeless battle for survival comes into its own. But it is the projection of a despairing mind in a universe estranged from the source of its existence.

In fact, this universe flows from the creative activity of a supreme consciousness of beauty and bliss. It is an emergent and interconnected process of self-shaped development, directed to the realisation of understanding, wisdom, creativity and relationship, and to eventual fulfilment in conscious relation to the infinite mind that is its source. It realises many distinctive forms of beauty and happiness, which otherwise could have no existence. It is also, necessarily, a universe in which struggle, competition and suffering exist, though many (not all) of its forms of suffering are necessary corollaries of the most complex and intense forms of happiness. It is a universe in which egoism can corrupt the good and bring suffering to the innocent, but in which that corruption is never without correction and that suffering is never without healing. It is a universe in which the purposes of God are destined to be realised, so that every suffering creature can be brought to experience an infinite personal good, and the positive potentialities of the universe can be brought to fruition.

With the development of rationally self-directing awareness, which on this planet reaches its highest form in humans, it becomes possible for creatures to choose forms of life that select egoistic pleasure over objective good. Such choices lead to an estranged form of life, when the presence of God is hidden and the power of God for good is weakened. The negative possibilities inherent in the creation of a cosmos of free, self-shaping creatures are realised, with the amplification of suffering and frustration for the whole planetary ecosystem. But, given the drive to complexity, integration and value in cosmic evolution, there is reason to hope that such negative possibilities will eventually be eliminated. Creatures will realise their positive potential, achieving conscious reintegration with the divine presence and power which ensures the fulfilment of the purpose for creation.

God will ensure that the goal of creation is realised, a goal of great value. God can also ensure that all finite persons will share in it, by taking the personal lives formed at a particular stage of cosmic evolution, and re-embodying them in supracosmic forms which make it possible for them to know and appreciate the whole history of the universe, as it is apprehended and preserved in the cosmic mind. They will also come to know God wholly, as supreme perfection, and revere the infinite divine beauty for its own sake.

If this is the way the world is, then it is a world that is freely and reasonably choosable by every free conscious agent. Such agents will see that this is a universe that realises a very great good – indeed, when one takes everlasting companionship with God into account, an infinite good. They will see that such a good, reached through a process of self-shaping attainment, could not exist other than in a universe in general like this. They will see that if there is to be a true communal life of freely self-shaping agents, one will have to share in the sufferings as well as in the joys of others. One may well suffer because of the egoistic and destructive acts of others, but one may also be able to offer one's own life out of compassion for those trapped in selfish desire, so that suffering will not be wholly futile. They will see, however, that they themselves may well choose a life of egoism, of greed, hatred and delusion, for the sake of the temporary goods it offers, even though it will bring suffering in its train, both for themselves and for others. And they will see that nevertheless they, each one of them, will be able to share in the infinite good that is the goal of creation.

Seeing these things, no rational agent would refuse to choose the existence of this universe. The innocent child does not suffer as a means to the greatest happiness of the greatest number. She suffers as a consequence, undesired and unintended by God, of the existence of a community of free, self-shaping agents, and because of the malevolent intentions of egoistic creatures. Nevertheless, she would choose the infinite good realised by this universe. Given the hope of endless life with God, she would choose to be a member of a community of persons that can find bliss in God for ever. She would choose to exist, in order to share in that good. In that short life on earth in which she suffers terribly, she would choose to offer her life out of compassion for all beings, if that was required and possible. If she was fully aware of the reality of God and of the hope of eternal happiness, she would choose life.

THE GOAL OF EVOLUTION

Richard Dawkins would probably protest that this simply begs all the important questions. How can we know that there is a God or a hope of eternal life? Is this not wishful thinking, invented precisely to make human life bearable in a blindly indifferent universe?

I hope it is clear by now that Dawkins' belief that this is a blindly indifferent universe is itself a piece of wishful thinking. The human wish it panders to is the wish to be independent, to be free of a prying, manipulating Big Daddy in the sky, to pursue one's own life without moral constraints, and to destroy traditional religious authorities. This is like the adolescent wish to be free of one's parental authority figures and to decide everything for oneself, whatever the outcome. It has a certain heroic bravado about it – one sees the 'free man' as the heroic quester for truth, throwing over reactionary constraints and facing up to the worst the universe can do, head unbowed and, certainly, knees unbent.

Perhaps such emotional factors are the things that drive our deepest beliefs; in which case, Dawkins and the theist are both seeing the universe in terms of their basic emotional stances. This is hardly a scientific issue. It goes deeper than that, to the very roots of human perception, motivation and action. Certainly, belief in God is, I have suggested, based on experiences, especially of key figures in human religious history, of a transcendent and empowering reality which transforms the human self from the bondage of egoism to the free pursuit of the good. And belief in eternal life is in turn based on the conviction, arising from experience of the perfection of the transcendent and of the quality of human relationship to it, that this process of transformation is able, in all sentient beings, to continue to its proper form of fulfilment. It is one thing for Dawkins to say that he has no such experience, and does not respect the testimony of those who say they have. It is quite another to say that evolutionary biology itself shows such convictions to be misplaced.

In fact, though evolutionary biology itself, as a scientific discipline, is silent on the subject of God's existence, it provides a quite remarkable array of data which strongly suggests the existence of at least an extremely wise and powerful designer. The universe does not look blind; on the contrary, it looks as if it has been contrived with the greatest intelligence. The universe does not look indifferent to value; it looks as if it is inherently directed to the realisation of goodness. Moreover, the theist believes that the creator is far from indifferent to the suffering that exists in creation. First, in knowing and experiencing all created reality, God suffers in and with all creatures. Second, in and through the process of cosmic causality, God constantly exercises a guiding influence, seeking to maximise good and eliminate evil, to the greatest extent compatible with preserving the autonomy of cosmic laws and the freedom of rational creatures. Third, God offers the possibility of eternal life to all rational creatures, and through them, the possibility of a proper fulfilment to all sentient beings.

The universe is not indifferent either to goodness or to the communities of sentient individuals who help to create and apprehend goodness. Neither genes nor memes are in themselves of intrinsic value. What is of value are the persons-in-relation who owe their physical existence to genes and consciously shape their futures by memes – or at least by ideas of truth, beauty and goodness. If that future includes everlasting relationship both to each other and to the infinite source of all being, then one can see how the hypothesis of God is one that provides the best explanation for this universe, with its complex mixture of good and evil. This is a universe that a perfect being could create, and one that points to such a being as its source.

The hypothesis of God must, of course, fit the facts. One cannot reasonably postulate a God who both wants to prevent all suffering and is able to do so simply by eliminating even the possibility of suffering, who could have created human beings perfect and fully formed, and who could have prevented all actual evil by a sheer act of divine will. But one *can* postulate a God who wills an intrinsic good of which suffering is a necessary condition or consequence, who wills that there should exist a universe containing communities of free, self-shaping rational creatures, whose being necessarily includes the possibility of suffering, and who expresses the divine nature in forms of relationship that move through suffering and conflict to reconciliation and fulfilment. The objections Richard Dawkins has are almost wholly objections to a naively imagined, anthropomorphic God, who is unreasonably, irrationally and blindly flattered and obeyed. The postulate of a supremely perfect God who is the source of all reality, and shapes it to share in supreme goodness, not only escapes his major objections, it is a strong implication of his beautifully expressed vision of a universe that begins with the confession that 'the living results of natural selection overwhelmingly impress us with the appearance of design'.[6] In fact, Dawkins writes, 'We animals are the most complicated and perfectly designed pieces of machinery in the known universe.'[7] I have tried to show that, although the design is more that of a dramatist or composer than of a watchmaker, and of a dramatist who is able to enter as participant into his (or her) own work, the appearance is not an illusion, but a pointer to the master intelligence that creates and guides the universe. The Darwinian worldview does not, in principle, 'solve the mystery of our existence'.[8] It is itself a speculative and dogmatic hypothesis which leaves most aspects of mind, value and intentionality wholly unexplained. The hypothesis of God is superior in explanatory power.

But, of course, really to believe in God is to have some experience of a being of transcendent power and value which is life-enhancing and value-transforming, and to trust the testimony of at least some of those who claim such experience to a pre-eminent degree. It is to experience a mystery beyond human comprehension, which sets limits to all human understanding, unless and insofar as it is empowered by the divine itself. Such an experience neither biology nor philosophy, nor indeed theology, can bring about. It may be, however, that the intelligible goal of the whole cosmic process is an entrance into the 'instructed ignorance' of love that lies beyond all proud claims to the omnipotence of human reason. There is no conflict between reason, the deepest understanding of the cosmos, and faith, the trusting response to the mystery of divine love. Together, they express the commitment to truth that should be the hallmark of science and the humility that should be the hallmark of faith. Only reason and faith together can bring humans and all sentient creatures to that maturity that is their proper form of life. Only then can the universe achieve that fully conscious relationship to its creator in which its created potentialities for good can find their proper fulfilment. That is the ultimate purpose of God and the goal of evolution.

NOTES

[1] See S. Weinberg, *Dreams of a Final Theory*, p. 105.
[2] Richard Dawkins, *River out of Eden*, p. 133.
[3] A powerful argument for this view is Paul Davies, *The Mind of God*.
[4] F. Dostoevsky, *The Brothers Karamazov*, pp. 274–87.
[5] There are many able defences of this view. A powerful one is A. Plantinga, *God, Freedom and Evil*.
[6] Dawkins, *The Blind Watchmaker*, p. 21.
[7] Dawkins, *The Selfish Gene*, p. xi.
[8] Dawkins, *The Blind Watchmaker*, p. xiv.

Bibliography

Anselm. *Proslogion* 2. M. J. Charlesworth (trans.). Notre Dame, Indiana, Notre Dame University Press, 1979.

Aquinas, T. *Summa Theologiae*, 1a, 2, 3. London, Eyre & Spottiswoode, 1964.

Atkins, P. *Creation Revisited*. Harmondsworth, Penguin, 1994.

Augustine. *De Genesi ad literam*, 4.26. J. H. Taylor (trans.). London, Newman Press, 1982.

Ayala, F. 'The Biological Roots of Morality' in *Biology and Philosophy* 2, 1987, pp. 235–52.

Barbour, I. *Religion in an Age of Science*. London, SCM Press, 1990.

Barrow, J. and Tipler, F. *The Anthropic Cosmological Principle*. Oxford, OUP, 1986.

Bartholomew, D. *God of Chance*. London, SCM Press, 1984.

Berkeley, G. *A Treatise Concerning the Principles of Human Knowledge*. New York, Scribner, 1929.

Campbell, D. T. 'Downward Causation in Hierarchically Organised Systems' in F. J. Ayala and T. Dobzhansky (eds.) *Studies in the Philosophy of Biology*. London, Macmillan, 1974, pp. 179–86.

Collier, J. and Stingl, M. 'Evolutionary Naturalism and the Objectivity of Morality' in *Biology and Philosophy* 8, 1993, pp. 47–60.

Darwin, C. *The Origin of Species by Means of Natural Selection* (first published 1859). Harmondsworth, Penguin, 1985.

Davies, P. *The Mind of God*. New York and London, Simon & Schuster, 1992.

Dawkins, R. *The Selfish Gene*. London, Granada Publishing, 1978.

Dawkins, R. *The Blind Watchmaker*. Harmondsworth, Penguin, 1991.

Dawkins, R. *River out of Eden*. London, Weidenfeld & Nicholson, 1995.

Dostoevsky, F. *The Brothers Karamazov*. D. Magarshack (trans.). Harmondsworth, Penguin, 1958.

Eccles, J. *Evolution of the Brain: Creation of the Self*. London, Routledge, 1989.

Eldredge, N. and Gould, S. J. 'Punctuated Equilibria: an Alternative to Phyletic Gradualism' in T. J. M. Schopf (ed.) *Models of Paleobiology*. San Francisco, Freeman, Cooper, 1972, pp. 82–115.

Ellis, G. 'The Theology of the Anthropic Principle' in C. J. Isham, R. Russell and N. Murphy (eds.) *Quantum Cosmology and the Laws of Nature*. Notre Dame, Indiana, Notre Dame University Press, 1993.

Feynman, R. *The Character of Physical Law*. Boston, MIT Press, 1965.

Goodwin, B. *How the Leopard Changed its Spots*. London, Weidenfeld & Nicholson, 1994.

Gould, S. *Wonderful Life: The Burgess Shale and the Nature of History*. London, Penguin, 1989.

Haji, I. 'Evolution, Altruism and the Prisoner's Dilemma' in *Biology and Philosophy* 7, 1992, pp. 161–76.

Hawking, S. *A Brief History of Time*. London, Bantam Press, 1989.

Houghton, J. 'New Ideas of Chaos in Physics' in *Science and Belief* 1, 1989.

Hume, D. *A Treatise of Human Nature*. Oxford, Oxford University Press, 1978.

Huxley, T. H. *Evolution and Ethics*. London, Pilot Press, 1949.

Isham, C. J. 'Quantum Theories of the Creation of the Universe' in C. J. Isham, R. J. Russell and N. Murphy (eds.) *Quantum Cosmology and the Laws of Nature*. Notre Dame, Indiana, Notre Dame University Press, 1993.

Isham, C. J., Russell, R. J. and Murphy, N. (eds.) *Quantum Cosmology and the Laws of Nature*. Notre Dame, Indiana, Notre Dame University Press, 1993.

Jaki, S. *The Road of Science and the Ways to God*. Chicago, University of Chicago Press, 1978.

Jaki, S. *Science and Creation*. Edinburgh, Scottish Academic Press, 1986.

Lack, D. *Evolutionary Theory and Christian Belief*. London, Methuen, 1961.

Leslie, J. *Value and Existence*. Oxford, Basil Blackwell, 1979.

Lovelock, J. *A New Look at Life on Earth*. Oxford, OUP, 1979.

McCoy, W. *Journal of Theoretical Biology*, 68.

Midgley, M. 'Gene-juggling' in *Philosophy* 54, pp. 439–58.

Midgley, M. 'Selfish Genes and Social Darwinism' in *Philosophy* 58, pp. 365–77.

Moore, G. E. *Principia Ethica*. Cambridge, CUP, 1951.

Patton, C. M. and Wheeler, J. A. 'Is Physics Legislated by Cosmogony?' in C. J. Isham, R. Penrose and D. W. Sciama (eds.) *Quantum Gravity*. Oxford, Clarendon Press, 1975.

Peacocke, A. *Theology for a Scientific Age*. London, SCM Press, 1993.

Penrose, R. *The Emperor's New Mind*. Oxford, OUP, 1989.

Penrose, R. *Shadows of the Mind*. Oxford, OUP, 1994.

Plantinga, A. *God, Freedom and Evil*. London, Allen & Unwin, 1974.

Plato. *Timaeus*. Harmondsworth, Penguin, 1965.

Polanyi, M. *Personal Knowledge*. London, Routledge, 1962.

Polkinghorne, J. *The Particle Play*. New York, Freeman, 1979.

Polkinghorne, J. *Science and Creation*. London, SPCK, 1988.

Poole, M. 'A Response to Richard Dawkins' in *Science and Christian Belief* 6, 1994.

Popper, K. R. *The Open Universe*. London, Hutchinson, 1982.

Prigogine, I. and Stengers, I. *Order out of Chaos*. London, Heinemann, 1984.

Quinton, A. *The Nature of Things*. London, Routledge, 1973.

Richards, R. J. *Darwin and the Emergence of Evolutionary Theories of Mind and Behaviour*. Chicago, University of Chicago Press, 1987.

Ruse, M. *Evolutionary Naturalism*. London and New York, Routledge, 1992.

Swinburne, R. *The Evolution of the Soul*. Oxford, Clarendon Press, 1986.

Taylor, R. *Action and Purpose*. New York, Prentice Hall, 1966.

Tipler, F. *The Physics of Immortality*. London, Macmillan, 1995.

Tryon, E. P. 'Is the Universe a Vacuum Fluctuation?' in J. Leslie (ed.) *Physical Cosmology and Philosophy*. New York, Macmillan, 1990.

Ward, K. *Defending the Soul*. Oxford, Oneworld, 1992.

Ward, K. *Religion and Creation*. Oxford, OUP, 1996.

Weinberg, S. *The First Three Minutes*. London, André Deutsch, 1977.

Weinberg, S. *Dreams of a Final Theory*. London, Vintage, 1993.

Wheeler, J. A. *Gravity and Spacetime*. New York, Scientific American Library, 1990.

Whitehead, A. N. *Science and the Modern World*. Cambridge, CUP, 1927.

Whitehead, A. N. *Process and Reality* (corr. and ed. D. R. Griffin and D. W. Sherbourne). New York, Free Press, 1978.

Wigner, E. 'The Unreasonable Effectiveness of Mathematics' in *Communications in Pure and Applied Mathematics* 13, 1990, pp. 1–14.

Index